P9-CDK-507

Thomas A Edison

Orange N.J.

If lost please return

TAE

AND THE RISE OF INNOVATION

LEONARD DeGRAAF

FOREWORD BY BILL GATES

Sterling Signature
NEW YORK

Sterling Signature
NEW YORK

An Imprint of Sterling Publishing
387 Park Avenue South
New York, NY 10016

ISBN 978-1-4027-6736-4

Distributed in Canada by Sterling Publishing
c/o Canadian Manda Group, 165 Dufferin Street
Toronto, Ontario, Canada M6K 3H6
Distributed in the United Kingdom by GMC Distribution Services
Castle Place, 166 High Street, Lewes, East Sussex, England BN7 1XU
Distributed in Australia by Capricorn Link (Australia) Pty. Ltd.
P.O. Box 704, Windsor, NSW 2756, Australia

For information about custom editions, special sales, and premium and corporate purchases, please contact Sterling Special Sales at 800-805-5489 or specialsales@sterlingpublishing.com.

Manufactured in China

2 4 6 8 10 9 7 5 3 1

www.sterlingpublishing.com

CONTENTS

EDISON'S
GREATEST MARVEL
THE VITASCO

TRADE
Thomas A. Edison
MARK
EDISON
PORTLAND
CEMENT
COMPANY
NEW VILLAGE
N.J.

FOREWORD

THERE'S NO QUESTION in my mind that one of America's greatest gifts to the world is our capacity for innovation. From light bulbs and telephones to vaccines and microprocessors, our inventions and ideas have improved the lives—and even saved the lives—of countless people around the globe.

In the pantheon of American innovation, Thomas Edison holds a unique place. He became a symbol of American ingenuity and the conviction that inspiration and perspiration could lead to remarkable things.

He certainly has been an inspiration to me in my career. I'm lucky enough to own a few pieces of Edison memorabilia, including his sketch of an idea for improving the incandescent light bulb and some papers on finding a substitute for rubber. Looking at this work, it's easy to see a creative mind continually trying to refine and improve his ideas.

Obviously, Edison's inventions were revolutionary. But as this book makes clear, the way he worked was also crucial for his success. For example, Edison consciously built on ideas from predecessors as well as contemporaries. And just as important, he assembled a team of people—engineers, chemists, mathematicians, and machinists—that he trusted and empowered to carry out his ideas. Names like Batchelor and Kruesi may not be famous today, but without their contributions, Edison might not be either.

Second, Edison was a very practical person. He learned early on that it wasn't enough to simply come up with great ideas in a vacuum; he had to invent things that people wanted.

That meant understanding the market, designing products that met his customers' needs, convincing his investors to support his ideas, and then promoting them. Edison didn't invent the light bulb; he invented the light bulb that worked, and the one that sold.

Finally, Edison recognized that inventions rarely come in a single flash of inspiration. You set a goal, measure progress using data, see what's working—and what isn't working—adjust your plan, and try again. This process can be very frustrating because it means running into a lot of dead ends. But each dead end tells you something useful. As Edison famously said, "I have not failed 10,000 times. I've successfully found 10,000 ways that will not work."

These lessons are just as true today as they were in Edison's time. Innovators still have to work in teams. (Although that's far easier to do today than at the turn of the twentieth century. Imagine what the Wizard of New Jersey's Menlo Park could have done with the tools coming out of California's Menlo Park.) Innovators still have to understand and solve real-world problems, and they still have to persevere for the long haul. Scientists run trial after trial to perfect a new vaccine. Co-workers at software companies debug each other's code.

While we've seen amazing advances in science and technology since Edison's day, these things have not changed. Thomas Edison remains a powerful exemplar of creativity, perseverance, and optimism. Even more than light bulbs and movie cameras, that may be his greatest legacy.

BILL GATES
Co-Chair, Bill & Melinda Gates Foundation
SEATTLE, WA

PREFACE

AS A MEASURE OF HOW Thomas Edison changed the world, consider this: When he was born in 1847, there were no industrial research laboratories, no phonographs, no motion picture cameras, and no electric power systems, let alone practical electric lights. In 1931, the year Edison died, the United States produced 320 million lightbulbs and consumed 110.4 million kilowatt-hours of electricity. Seventy-five million Americans attended the movies each week, spending $719 million ($10.6 billion today) at the box office.[1]

In the year of Edison's death, the *New York Times* estimated the value of the industries based on his inventions at more than $15 billion. His inventions made the modern age possible. Without improved telegraph, telephone, and electric power systems and the ability to record, store, and transmit sound and images, there would be no Internet or computers.

From the 1870s through the 1920s, Edison's laboratories combined knowledge, resources, and talented collaborators to turn ideas into commercial products. His laboratory workers invented the phonograph, a practical incandescent electric lighting system, and the motion picture camera. They also developed the nickel-iron storage battery, machinery for processing iron ore and manufacturing Portland cement, and a system for constructing molded cement houses. In his last experimental project, Edison created a process for extracting rubber from goldenrod, a flowering plant considered by most to be a weed.

Over the course of his long career, Edison organized and managed dozens of manufacturing and marketing companies. In the process of adopting new production and sales strategies, he helped create a mass consumer market in the late nineteenth and early twentieth centuries. Edison was also one of the first business leaders to brand himself, paving the way for Walt Disney, Steve Jobs, and other modern entrepreneurs.

Foreshadowing closer government-business relations in the twentieth century, Edison was a vocal proponent of military and industrial preparedness during the First World War. He also conducted research for the U.S. Navy during the war and served as president of the Naval Consulting Board, a group of civilian technical experts established to advise the navy on ideas for inventions.

Edison helped change the way technologies were developed. Benefiting from advances in science and the multidisciplinary labors of chemists, engineers, mathematicians, and other trained professionals, Edison's laboratories introduced new products on a regular basis. Invention shifted from talented individuals working alone to organized groups working in laboratories established specifically for industrial research and development.

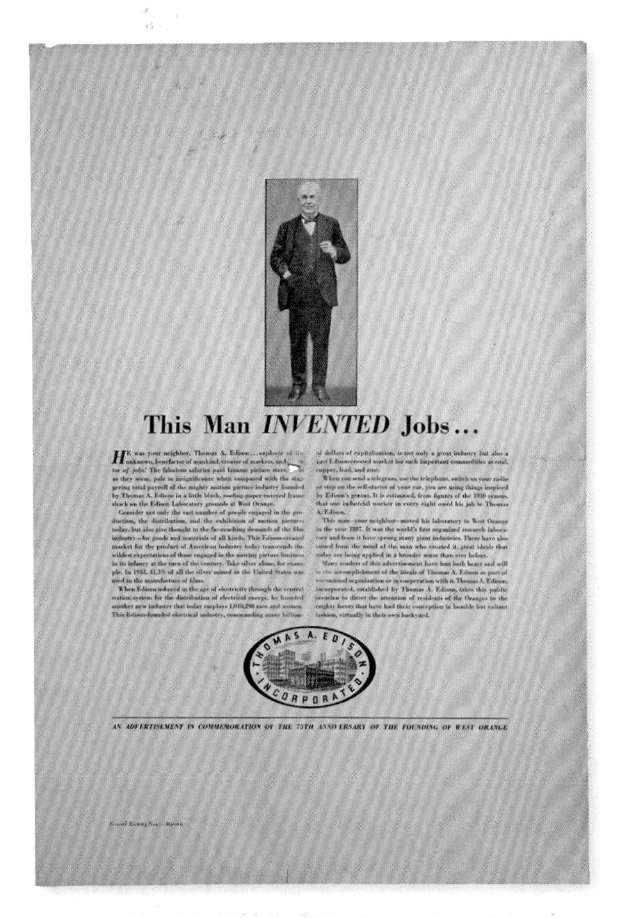

Edison's Menlo Park and West Orange labs created new industries, including electric lighting and power, sound recording, and motion pictures. Thomas A. Edison, Inc., promoted this economic legacy in the 1940s and 1950s.

A century before the modern globalization of the world's economy, Edison operated on an international scale. He manufactured and marketed his inventions in Europe, North and South America, and Asia. He relied on globally sourced raw materials and skilled workers. Ideas and concepts generated by an international community of scientists and researchers influenced his work, and, in turn, a global public eagerly awaited his "latest invention."

Edison's experience as an innovator is as relevant today as it was one hundred years ago. Edison devoted considerable attention to the questions all innovators face in modern times:

Edison's desk in the library of the West Orange lab. The cubbyholes reflect the diversity of his inventions and business interests.

Which products should I develop? How should those products be designed, manufactured, and marketed? How do I raise money to support research and development? How do I respond to competition and changing markets? Knowing Edison's response to these questions brings us closer to understanding the nature of technological innovation and creativity.

Edison stands out as an innovator, not because he always succeeded, but because of the scope and range of his interests. His ability to pursue research in diverse fields and draw upon past experiences to solve new problems were among his greatest strengths. The cubbyhole in his West Orange laboratory desk marked "New Things" reminds us of his irrepressible interest in the next big idea. If he were alive today, he would be on the cutting edge of innovation.

THE HISTORIC SITES AND MEMORIALS preserved for our benefit are a tangible legacy of his work. These include his Milan, Ohio, birthplace; the Port Huron, Michigan, train depot where he worked during the Civil War; and the house in Louisville, Kentucky, where he lived in 1866 during his years as an itinerant telegrapher. Henry Ford reconstructed Edison's Menlo Park laboratory in the late 1920s at Greenfield Village in Dearborn, Michigan, and Edison's former electric light associates established a memorial tower on the site of the original Menlo Park lab in the 1930s. Edison's widow, Mina Miller Edison, donated their Fort Myers, Florida, winter home to the city of Fort Myers.

Thomas Edison National Historical Park preserves and interprets his West Orange laboratory and nearby home, Glenmont, where he lived and worked for the last forty-five years of his life. The park's museum collection is a rich trove of 400,000 artifacts, 48,000 sound recordings, 60,000 historic photographs, and more than five million archival documents. Most of the documents and photographs in this book come from the park's collections and are presented here to make them more accessible.

The Charles Edison Fund and Edison Innovation Foundation are proud to join with Sterling Publishing and the National Park Service to present this book on Edison, the Innovator. The Fund represents the philanthropic efforts of Thomas's son, Charles, who followed in the benevolent aims of his father and mother, Mina Miller Edison. Charles served as governor of New Jersey during the Second World War and, as secretary of the navy under President Franklin D. Roosevelt, championed the Iowa-class battleship and PT boat. The fund promotes the legacy of Thomas Edison and supports medical research, educational outreach, and historical preservation at Thomas Edison National Historical Park.

The Edison Innovation Foundation is a not-for-profit organization that supports the Edison legacy and encourages students to embrace careers in science and technology. The Foundation has a partnership agreement with the National Park Service to raise funds for Thomas Edison National Historical Park. Fund-raising efforts over the last five years have raised in excess of $20 million for the conservation, renovation, and preservation of the park's significant collection of historical buildings and artifacts, making them accessible to future generations.

You can contact the Foundation at Thomasedison.org and through our Facebook page. As you travel with Thomas Edison on his amazing journey, enjoy this book, and remember his most famous quote: "There is a better way to do it. *Find it!*"

JOHN P. KEEGAN
President & CEO
CHARLES EDISON FUND
EDISON INNOVATION FOUNDATION

INTRODUCTION

THERE ARE TWO THOMAS EDISONS: the mythic, larger-than-life "Wizard of Menlo Park"—a tireless heroic inventor with an inexhaustible supply of ideas who gave us light, sound, and moving pictures—and the Innovator who spent his life solving technical problems in shops and laboratories and creating companies to manufacture and market new technologies.

One Edison could not have existed without the other. The Wizard inspired faith in financial backers, business partners, employees, and consumers, providing the Innovator with the resources and support he needed. The Wizard was based on a solid record of accomplishment; without the successes of the Innovator, no one would have paid any attention to the Wizard.

Both personas are significant. Edison's fame reflected American faith in technological progress and reaffirmed a compelling national narrative: a young man with no family connections, no wealth, and little formal education rises through hard work, ingenuity, and perhaps a little luck to become the preeminent inventor of his age. Edison's story reveals a different picture: a talented inventor and entrepreneur who worked within and took advantage of broader social, cultural, and institutional forces that ushered in the modern age of high-tech industrial capitalism, globalization, and consumerism.

The Wizard often overshadowed the Innovator, contributing to misleading assumptions about Edison. Maurice Holland, the director of the National Research Council's Division of

OPPOSITE: Edison examining a disc phonograph record at the West Orange lab, June 1921.

Engineering and Industrial Research, encountered both Edisons during a 1927 visit to the West Orange laboratory. Holland scheduled five days to study Edison's method of invention. He wanted to know if Edison's success could be attributed to his personal genius or the organization of his laboratory.

Edison welcomed Holland cordially, invited him to stay as long as he liked, and told him he did not need five days. "There is no organization; I am the organization," Edison proclaimed. When Holland asked Edison how many experiments were in progress, the Wizard answered, "I haven't any idea. I have enough ideas to keep the laboratory busy for years and break the Bank of England." Holland described the laboratory as "a number of rambling buildings without any definite plan, obviously a product of evolution" and observed "elaborate records, detailed cost accounting, scheduling, requisitions and myriad devices considered necessary under the heading 'Business Organization' has no place here when judged by the practical standards of accomplished results."

The Innovator, of course, knew exactly how many experiments were under way in the laboratory and precisely what his experimenters were doing. His businesses were organized; the laboratory's physical layout was planned; and—as for the lack of financial records—187 linear feet of accounting records and 216 linear feet of requisition records survive in Edison's archive.

Holland got some things right. He accurately observed Edison's reliance on detailed experimental records and recognized his team approach to invention. After what he called "persistent questioning," Holland coaxed from Edison his method of experimentation:

> First state the problem clearly . . . formulate it, then write down every conceivable means of solution, including, as he expressed it 'the damned fool ideas, for they sometimes work,' build an experimental model to prove the principle, put it to every conceivable test encountered in actual service, standardize it for manufacture, turn it over to production experts for commercial production.

This is a fair enough description of how Edison worked, but Holland said nothing about how much time Edison spent on administrative, manufacturing, and marketing problems.

Holland's report illustrates the difficulty of sorting Edison myth from reality. Edison contributed to this problem by reducing invention to witty sayings like "Genius is 1%

WHAT IS INVENTION?

WHAT WERE HIS QUALITIES OF GREATNESS?

IMAGINATION AND OBSERVATION

An example of Edison's fruitful imagination and acute observation is the phonograph, his most original invention. Experimenting on his automatic telegraph repeater, he heard a sound as the metal point passed in and out of perforations on the whirling disk. That night he wrote, "There is no doubt that I shall be able to store up and reproduce automatically at any future time the human voice perfectly." This thought, completely new to mankind, became a fact just three weeks later with the invention of the phonograph.

KNOWLEDGE AND MEMORY

Edison was a prodigious and omnivorous reader, early becoming so adept that he read by paragraphs rather than by words. With almost no formal education he became an expert in many fields through his habit of reading everything that had been written on the subject before starting a new project. To illustrate his phenomenal memory, he once spent a day at his cement plant in New Village, N. J. without taking any notes, returned to his Orange Laboratory, and compiled a list of 1800 detailed suggestions to the manager.

1% INSPIRATION, 99% PERSPIRATION

OPTIMISM AND HUMOR

That Edison was never discouraged was a source of perpetual amazement to his assistants. One day during the development of his storage battery, an epic work involving 50,000 experiments, when asked if he was not discouraged, Edison replied, "No, we now know several thousand things that won't work." Always buoyed up by his optimism and humor, he could find time for fun even in the midst of the deepest problems, and an admonishment could very frequently be forestalled by mollifying "the old man" with a good new joke.

TEAMWORK AND UNSELFISHNESS

Realizing early that manufacturing took up too much of his time, Edison started his custom of turning over the manufacturing to others. By selling the patents and interests in these ventures he financed more research and to increase the output of inventions, used assistants to carry out detailed experiments under his direction. Thus began organized research. His unselfishness was in concentrating his genius on what he could do best – invention and research – and turning over to others what they could do best.

AND PERSISTANCE DESPITE FAILURES

VOTE RECORDER

The parliamentary vote recorder, Edison's first invention, was refused by Congress because delaying was a privilege of the minority. This taught him not to invent anything unwanted.

ORE SEPARATOR

Edison's magnetic ore separator utilized low grade iron ore. $2,000,000, spent on it went for nothing when rich deposits were discovered. He had to convert it to cement making.

ELECTRIC LIGHT

In search of a better bamboo lamp filament, Edison sent experts around the world. The better bamboo, when found, was already outmoded by his newly developed artificial filament.

STORAGE BATTERY

Ten years of arduous work and fifty thousand experiments finally enabled Edison to perfect the iron-nickel-alkaline storage battery, a battery that wouldn't be self-destructive.

Thomas A. Edison, Inc., combined the qualities of Edison the Innovator (scientific knowledge, teamwork) with the Wizard (optimism, humor, persistence) in this 1940s magazine ad.

inspiration, 99% perspiration"—folksy axioms that entertained newspaper readers but did nothing to explain how he operated. Edison attributed his success to hard work and what a San Francisco newspaper in 1898 called his "dogged persistence." Edison's legendary capacity for hard work, curiosity, and entrepreneurial drive are relevant, but these characterizations oversimplify a complex process.

Compounding the mystery are misleading assertions about Edison's role as an innovator. Henry Ford, a close friend for the last two decades of Edison's life, called him "the world's greatest inventor but worst businessman." In his 1985 book *Innovation and Entrepreneurship*, management expert Peter Drucker erroneously claimed that Edison "so totally mismanaged the businesses he started that he had to be removed from every one of them to save it" and that Edison's principal ambition was to become a "tycoon." This view of Edison as a poor business manager is often declared without the benefit of research on the organization and management of Edison's companies.

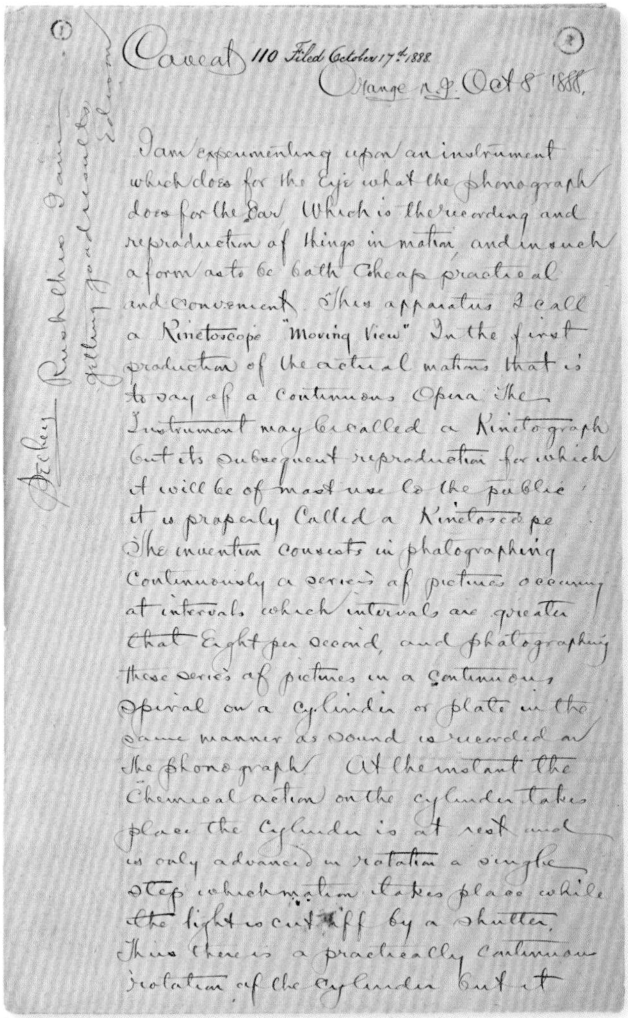
Caveat 110 Filed October 17th 1888.
Orange N.J. Oct 8 1888.

I am experimenting upon an instrument which does for the Eye what the phonograph does for the Ear, which is the recording and reproduction of things in motion, and in such a form as to be both Cheap practical and convenient. This apparatus I call a Kinetoscope "Moving View". In the first production of the actual motions that is to say of a continuous Opera the Instrument may be called a Kinetograph but its subsequent reproduction for which it will be of most use to the public it is properly called a Kinetoscope. The invention consists in photographing continuously a series of pictures occurring at intervals which intervals are greater that Eight per second, and photographing these series of pictures in a continuous spiral on a cylinder or plate in the same manner as sound is recorded on the phonograph. At the instant the chemical action on the cylinder takes place the cylinder is at rest and is only advanced in rotation a single step which motion takes place while the light is cut off by a shutter. Thus there is a practically continuous rotation of the cylinder but it

First page of Edison's October 1888 patent caveat, describing his plan to invent a motion picture camera.

With millions of documents, thousands of historic photographs and artifacts, and several carefully preserved historic sites and museums in the United States, we have the resources for a fresh examination of Edison. This book traces Edison's long career, from his formative years as a telegraph inventor in the early 1870s, to his first major research laboratory at Menlo Park, to the West Orange laboratory where Edison worked during the last forty-five years of his life.

Following a brief review of Edison's childhood and youth in this introduction, the first chapter examines his first steps toward becoming an inventor in the early 1870s. Subsequent chapters focus on the Menlo Park and West Orange laboratories as organized research facilities or on specific technologies, including the phonograph, the electric light, motion pictures, ore milling, Portland cement, the storage battery, and rubber. Chapters also describe Edison's transition from Menlo Park to West Orange and his role as an innovator during the First World War. A concluding chapter discusses Edison's final illness and death and his memorialization through the preservation of historic sites and museum collections associated with his life.

Sidebars introduce lesser-known Edison inventions or products, including the electric pen, the electric railroad, the talking doll, X-ray equipment, and the Edicraft appliances. Sidebars also explore Edison's laboratory notebooks and the 1880s "battle of the currents."

In reframing Edison, this book argues that innovation is not simply a linear "idea-to-product" process, where inventors take an idea, build a prototype or working model, put the model into production, and sell the manufactured articles to consumers. Instead, a close examination of Edison's innovation methods reveals a social process involving the interaction of inventors, manufacturers, marketers, consumers, and others answering important questions about which technologies society should develop and how they should be designed, produced, marketed, and consumed—answers shaped by the goals,

Edison in June 1881.

Edison relied on the technical and scientific reputation of his laboratory to market his Portland cement.

values, and assumptions of the people involved in the process. From this perspective, Edison's inventions reflected the evolution of broader societal conditions, the goals of his investors and business associates, and the information and resources available to him, in addition to his own perceptions of what society needed or wanted.

THE CHARACTER TRAITS that defined Edison emerged in his childhood and youth. Edison was born in Milan, Ohio, on February 11, 1847, the youngest of Samuel and Nancy Elliott Edison's seven children. Milan was connected to Lake Erie by a short canal, which helped the village become a prosperous shipbuilder and grain port. Milan's thriving economy allowed Edison's father—a carpenter, shingle maker, and land speculator—to build a sturdy brick house on the bank of the canal. Edison left Milan when he was seven years old, but he recalled the grain elevators along the canal and the busy shipyard. He also remembered the town square filled with wagon teams delivering barrel staves and covered wagons staged in front of his house, preparing for their journey to the California gold fields.

Milan's economy declined the early 1850s after the opening of a railroad between Mansfield, Ohio, and the Lake Erie port of Sandusky allowed grain shippers to bypass the canal. Samuel Edison moved his family to Port Huron, Michigan, in the spring of 1854 to find better opportunities. In Port Huron, Samuel sold lumber, speculated in real estate, and built and operated an observation tower that offered tourists views of Lake Huron. Samuel also worked a ten-acre vegetable farm. Edison remembered working on this farm, planting corn, onions, and other vegetables. With a horse and wagon, he helped Samuel sell the crops door-to-door.

Edison may have learned entrepreneurship from his father, but his mother, a former schoolteacher, was largely responsible for his early education. Edison attended school for less than a year. There is no evidence that a teacher dismissed Edison because of a learning disability; rather, the family's limited resources may have prevented additional formal schooling. It was not uncommon for boys of Edison's age and social class in the mid-nineteenth century to receive the basics of reading, writing, and simple arithmetic and then to take jobs to help supplement family incomes.

Edison credited his mother with teaching him how to read. As he told New Jersey grammar school students in 1912, "My mother taught me how to read good books quickly and correctly, and as this opened up a great world in literature, I have always been very thankful for this early training." As an adult, Edison was a voracious reader, and his ability to read and absorb large quantities of printed information contributed significantly to his success.

"IT IS TOO MUCH THE FASHION TO ATTRIBUTE ALL INVENTIONS TO ACCIDENT, AND A GREAT DEAL OF NONSENSE IS TALKED ON THAT SCORE."

Nancy, a devout Methodist, may have taught Edison to read from the family Bible. Edison also had access to Samuel's library of political tracts, which included works by Thomas Paine. Beyond basic reading and writing, Edison demonstrated an early interest in science, reading elementary

LEFT: Edison's father, Samuel Edison Jr. RIGHT: Edison's mother, Nancy Elliott Edison.

textbooks like Richard Green Parker's *First Lessons in Natural Philosophy* (1859). Archeological studies of Edison's Port Huron home have uncovered remnants of a chemistry set.

The railroad and telegraph—two technologies that transformed nineteenth-century America—gave Edison his first significant employment opportunities. In 1859 Nancy gave him permission to work as a newsboy for the Grand Trunk Railway. Each morning, twelve-year-old Edison boarded a 7:00 a.m. Detroit-bound train to sell newspapers, magazines, candy, and fruit. The train returned to Port Huron at 9:00 p.m. Soon, Edison employed two Port Huron boys to operate stores selling magazines, vegetables, and butter. He stocked the stores with produce purchased in Detroit at wholesale prices or from farmers along the railroad. To save on shipping costs, Edison used unoccupied space in the baggage car and enlisted the cooperation of railroad workers by selling discounted butter and vegetables to their wives. Edison expanded his news business by hiring boys to sell newspapers on other trains.

In early 1862 Edison purchased a secondhand printing press and began publishing a newspaper, the *Weekly Herald*, in the train's baggage car. For a monthly subscription of eight cents, the newspaper provided readers with local news, train schedules, birth announcements, advertisements, and egg, butter, and vegetable prices. In his spare time, Edison experimented with chemicals in the baggage car, until a jar of phosphorus caught fire and the angry conductor, Alexander Stephenson, tossed the printing press and chemicals off the train.

Edison noticed a loss of hearing at an early age. No one knows the cause of Edison's deafness, but he later blamed it on conductor Stephenson who, depending on the account, either boxed Edison's ears or lifted him by the ears onto a railcar. A doctor who examined Edison later in his life believed that the hearing loss was caused by a congenital defect, but scarlet

Edison's birthplace, Milan, Ohio.

fever, which Edison contracted soon after the family moved to Port Huron, may have contributed to the problem. Whatever the cause, Edison claimed that his hearing loss did not seriously hinder his work because it allowed him to concentrate and disregard extraneous noise.

An act of bravery in the fall of 1862 gave Edison a life-altering opportunity. While standing on the Mount Clemens, Michigan, station platform, Edison saw the stationmaster's three-year-old son in the path of an oncoming railcar. As Edison's secretary later recalled, "Springing to his assistance, Edison succeeded in getting the boy off the track a few seconds before he would have been crushed." In gratitude, stationmaster James Mackenzie fed Edison dinner for three months and, more important, taught him telegraphy.

In the winter of 1863 the fifteen-year-old Edison took his first job sending and receiving telegraph messages at a jewelry store in Port Huron. The following summer he worked as a telegraph operator for the Grand Trunk Railway in Stratford Junction, Ontario, and between 1864 and 1867 he was employed as an itinerant telegrapher in

TOP: The young Edison read about science in textbooks like Richard Green Parker's *First Lessons in Natural Philosophy*. BOTTOM: Edison at age fourteen.

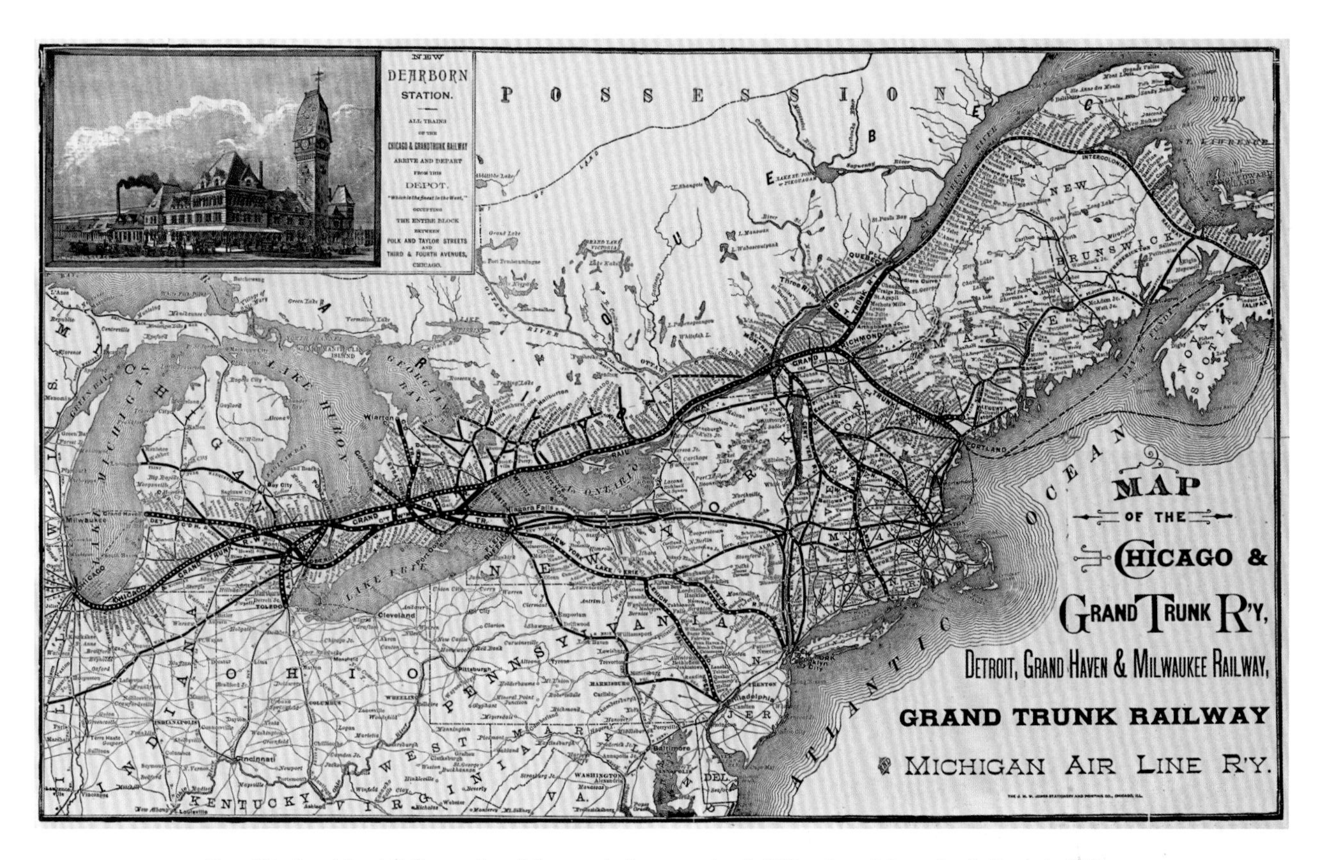

Map of the Grand Trunk Railway, where Edison worked as a newsboy in 1859 and as a telegrapher in the early 1860s.

cities throughout the Midwest and South, including Cincinnati, Fort Wayne, Indianapolis, Memphis, and Louisville.

Most telegraphers in the post–Civil War 1860s were young, unattached men who moved from town to town, looking for positions in telegraph offices. They were part of a community of skilled operators who acquired a reputation for leading carefree, dissolute lives. Operators were expected to have a basic understanding of the technology so that they could maintain, repair, and adjust equipment, including the chemical batteries that powered the system. As a telegrapher, Edison acquired these skills, but he was part of a smaller group of operators who wanted to further their education through self-study and to design improved telegraph equipment.

Because Edison was better at receiving telegraph messages (listening to the clicks of an incoming signal and transcribing the message on paper) than sending (tapping outgoing

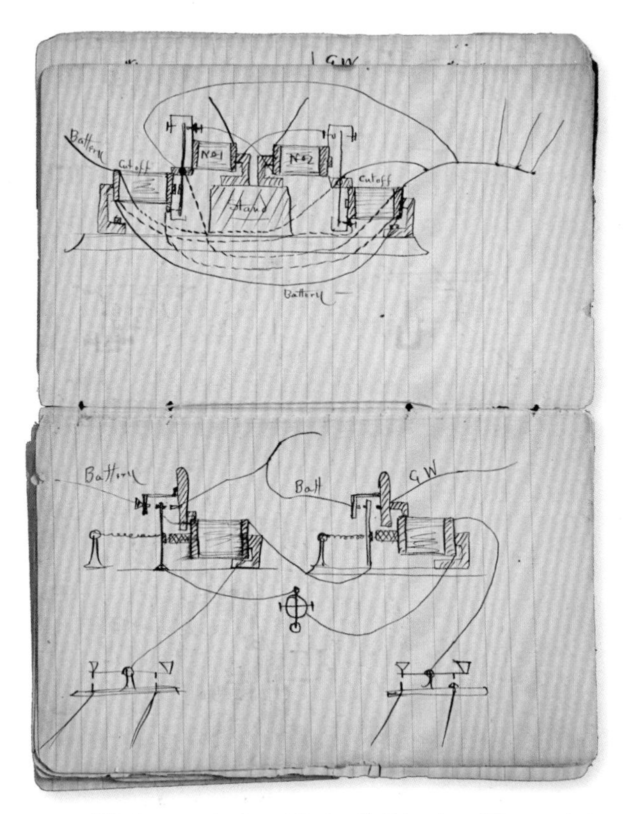

TOP: A panoramic view of Cincinnati, Ohio, where Edison worked as a telegraph operator for Western Union in 1865 and 1867. BOTTOM: Edison's early technical education included drawing designs for telegraph circuits and relays, as seen in these pages of a pocket notebook from the late 1860s.

messages on a telegraph key), he worked the night shift, when press services transmitted long newspaper copy. Working nights gave Edison ample time to read, study the technical literature, and experiment. In an early notebook Edison kept when he worked in Cincinnati in 1867, he sketched ideas for improving telegraph equipment, including designs for new relays that would increase the strength of incoming telegraph messages; repeaters that would allow the transmission of messages over longer distances; and multiple telegraph circuits, designed to send more than one message over the same wire.

With spare money, Edison purchased books, tools, and equipment to conduct experiments. Among the technical books Edison read in the late 1860s were Michael Faraday's *Experimental Researches in Electricity* (1855) and Dionysius Lardner's *Electric Telegraph* (1867). Edison's frequent book purchases nearly led to his being shot in 1866

in Louisville, where he bought fifty volumes of old *North American Reviews* at a junk auction. As Edison recalled:

> One morning after getting through the press I took 10 volumes on my shoulder & started for home. It was rather dark & while nearing home which was a room above a saloon I heard a shot & stopped. A policeman ran up & grabbed me by the throat. Fortunately I knew him. He had yelled but I being rather deaf did not hear . . . he supposed I had stolen the books.

Edison left Louisville and returned home to Port Huron in the fall of 1867 for an extended visit with his family. After spending a few months in Port Huron, he became restless and asked a telegrapher friend who worked in Boston if there were any jobs for him. The friend told Edison to come immediately.

EXPERI÷MENTAL RESEARCHES IN·ELEC-TRICITY *by* MICHAEL FARADAY D.C.L. & F.R.S.

LONDON: PUBLISHED by J.M.DENT & SONS L^TD^ AND IN NEW YORK BY E.P.DUTTON & CO

Edison's understanding of electricity and magnetism, crucial to his success as a telegraph inventor, was influenced by the research of British scientist Michael Faraday. The work of Faraday, inventor of the electric motor, laid the scientific foundation for electrical technologies in the nineteenth century.

1

THE EDUCATION OF AN INNOVATOR

"ALL MY LIFE I HAVE BEEN A COMMERCIAL INVENTOR. I HAVE NEVER DABBLED IN ANYTHING THAT WAS NOT USEFUL."

BOSTON WAS A SUPPORTIVE ENVIRONMENT for aspiring young inventors like Edison. It had been a major telegraph equipment manufacturer since the 1850s and was a leading developer of urban telegraph systems. Western Union and the Franklin Telegraph Co. had offices in Boston. Several influential telegraph inventors, including Moses Farmer and Joseph Stearns, lived there. Boston also had entrepreneurs with money to invest in machine shops that gave independent inventors access to machine tools, equipment, and skilled machinists in order to develop and test new ideas.

After Edison arrived in Boston in the spring of 1868, Western Union hired him to receive night press reports from New York in its main office. This gave Edison free time during the day to visit machine shops, experiment on new ideas, and write articles on technical subjects for the telegraph industry journal, the *Telegrapher*.

Edison came to Boston at an auspicious time. Between 1867 and 1870 the telegraph system grew from 85,000 to 112,000 miles of wire, and annual message volume increased from 5,879,000 to 9,158,000. Both wire mileage and message traffic continued to rise steadily through the 1870s. Growing American cities demanded specialized telegraph systems, including fire and police call systems, private line telegraphs that allowed communication between residences and offices, and reporting services that provided financial and commodity price information to subscribers from central locations. Telegraph companies needed new technologies to help meet the demand for their services and maintain their competiveness in a growing and rapidly changing business. Through self-education and his experiences as an itinerant telegrapher in the late 1860s, Edison had devised ideas for improving telegraph technology. Bos-

PAGE 1: Hand tools in the West Orange lab in the West Orange lab heavy machine shop. Edison's knowledge of machine shops and access to skilled machinists heped launch his career. ABOVE: Boston office of the Western Union Telegraph Co. Boston's telegraph companies and machine shops offered a supportive environment for the ambitious young Edison.

ton provided him with resources and support to develop those ideas and launch his career as a professional inventor.

Edison in 1884.

IT DID NOT TAKE Edison long to find investors. Telegrapher Dewitt C. Roberts, who gave Edison money to manufacture a stock printer in April 1868, was an early financial backer. Ebenezer B. Welch, a Boston merchant and director of the Franklin Telegraph Co., supported Edison's experiments on a fire alarm telegraph and a double transmitter—a device designed to send two telegraph messages at the same time.

Finding investors was not always easy, though. Edison could not secure funding for another early idea: a facsimile telegraph that transmitted Chinese characters. As he wrote a friend in September 1868, "It will probably be several months before I will be able to bring it out as experiments are rather Costly and there is a scarcity of funds."

Edison signed the application for his first patented invention, an electrographic vote recorder, on October 13, 1868. The vote recorder allowed legislators to transmit votes on bills to a central recorder, which automatically tabulated the results of the roll call. "This contrivance would save several hours of public time every day in the session, and I thought my fortune was made," Edison later recalled. When he took the vote recorder to Washington, however, he learned that lawmakers were not interested in the machine. "Young man," he was told, "that is just what we do not want. Your invention would destroy the only hope the minority would have of influencing legislation." Claims that Edison, after this experience, vowed to invent only things people wanted may be apocryphal, but it was a valuable lesson for him on the importance of knowing the market for new products.

On January 21, 1869, Edison signed an agreement with two Boston merchants, Joel Hills and William E. Plummer, who helped Edison patent an improved printing telegraph. Better

Edison's first patented invention, the electromagnetic vote recorder.

known as stock tickers, printing telegraphs transmitted stock prices to bankers and stockbrokers from a central office. Receiving instruments in brokers' offices printed the information on strips of paper. Edison's stock ticker was simpler than existing printers, using only one wire instead of three. With the support of Hills and Plummer, Edison opened a stock quotation business that served twenty-five subscribers.

This business arrangement allowed Edison to resign from Western Union and announce his decision to become a full-time inventor. On January 30, 1869, he published a notice in the *Telegrapher*: "Mr. T.A. Edison has resigned his situation in the Western Union office, Boston, Mass., and will devote his time to bringing out his inventions." For the next six years, he invented and manufactured telegraph systems. In terms of patent applications, these were productive years: between 1869 and 1875 Edison submitted 110 patent applications to the U.S. Patent Office. During this formative period, he learned how to manage shops, work with business partners, secure patent protection for his inventions, and recruit talented workers. Most important, Edison developed relationships with prominent telegraph industry managers and engineers, who provided financial, legal, business, and technical support and who influenced his research agenda. In short, Edison acquired the skills and experience that allowed him to become an effective innovator at Menlo Park and West Orange.

In April 1869 Edison tested his double transmitter on a telegraph line between New York City and Rochester, with the assistance of Franklin Pope, an influential electrical engineer, former *Telegrapher* editor, and superintendent of the Gold and Stock Telegraph Co. The double transmitter was an early effort by Edison to send more than one telegraph message over a single wire, but the test over the 400-mile-long circuit failed. At that point, Edison decided to move to New York because, as he told a friend, "It is useless for me to lay around or come to

Boston, for I cannot make any money there." Edison's Boston supporters lacked the funds to improve Edison's stock printer, double transmitter, and a dial telegraph or magnetograph—a private telegraph system that allowed businesses to communicate with distant factories. Edison hoped that investors in New York, the nation's leading financial and commercial center, would have the resources to support his research.

Edison's work came to the attention of the telegraph industry officials in New York, particularly Samuel S. Laws, the president of the Gold and Stock Reporting Telegraph Co., which transmitted gold price information to New York banks and commodities brokers. When Pope resigned as superintendent of Gold and Stock in August, Laws appointed Edison as his successor. Edison lost this position within four weeks when Gold and Stock merged with a competitor, so the young inventor moved to Elizabeth, New Jersey, where he established an electrical consulting business with Pope and *Telegrapher* editor James Ashley.

During the early 1870s several prominent telegraph industry managers supported and influenced Edison's inventions. Marshall Lefferts, a telegraph engineer and president of the Gold and Stock Telegraph Co., was Edison's first significant mentor. Lefferts had been involved in the telegraph business since the late 1840s and, during the Civil War, was general manager of the American Telegraph Co. In the late 1860s he was the superintendent of Western Union's Commercial News Department. Under Lefferts, Gold and Stock held a monopoly on the transmission of New York Stock Exchange information—a position Lefferts was determined to maintain by aggressively controlling printing telegraph technology. Lefferts recognized that Edison's technical abilities would help his company develop equipment to maintain its hold on the stock reporting business.

Marshall Lefferts, president of the Gold and Stock Telegraph Co., was Edison's first important business mentor.

On February 10, 1870, Edison signed two contracts with Gold and Stock. In one contract, the company agreed to pay Edison $7,000 ($124,000 today) for inventing an improved stock printer. In the other contract, Gold and Stock would pay Edison $3,000 ($53,300 today) for inventing a "simple, reliable, practical" facsimile telegraph to transmit written characters. Gold and Stock also agreed to pay Edison's rent for experimental space, up to $400 ($7,110 today) for tools and machinery, and the salary of a research assistant (to be hired by Edison) for six months. The agreements with Gold and Stock enabled Edison to form a partnership with machinist William Unger and open his first major shop, the Newark Telegraph Works, in Newark, New Jersey.

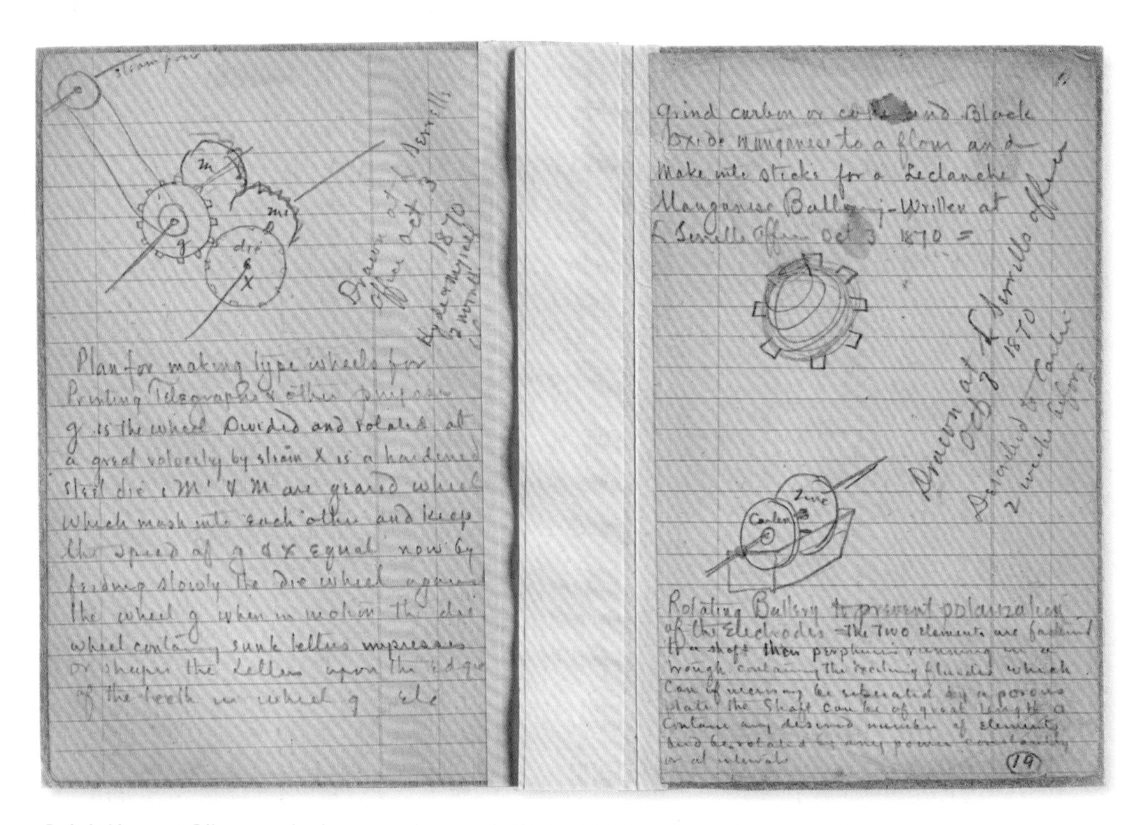

Early in his career, Edison vowed to keep a complete record of his experiments. This early 1870s notebook contains his notes and sketches for telegraph instruments.

Marshall Lefferts introduced Edison to Daniel Craig, the general agent of the National Telegraph Co. and a former president of the Associated Press who had invested in the automatic telegraph system of English inventor George Little. Automatic telegraphs allowed the transmission of messages without the need for an operator to click Morse code on a telegraph key. The dots and dashes were instead coded on strips of paper with a series of punched holes. When the paper passed through a transmitter, the holes allowed contacts to briefly close the circuit, thereby sending the message. At the receiving end, messages were automatically recorded on paper. The automatic telegraph promised more efficient and less costly message transmission. Instead of experienced operators, telegraph companies could employ less-skilled workers to prepare messages by using perforators and feeding the paper strips into the transmitter.

Lefferts and Craig wanted Edison to improve components of Little's system. In August 1870 Edison agreed to invent an improved perforator, for which he would receive $1,300 in cash and $3,700 in National Telegraph Co. stock ($23,100 and $65,000 today, respectively).

Little's system had technical defects and required further development to make it practical. Consequently, Craig enlisted the financial assistance of George Harrington, an entrepreneur who had been an assistant secretary of the treasury in the Lincoln administration and U.S. minister to Switzerland from 1865 to 1879. In November 1870 Craig and Harrington organized the Automatic Telegraph Co. to develop and promote automatic telegraph systems. Harrington, who became president of the Automatic Telegraph Co., also encouraged Edison to design his own automatic system and formed a partnership with him in October 1870 to organize the American Telegraph Works, which manufactured components for the Little system and printing

Pope-Edison telegraph printer. In 1869 and 1870, Edison shared three patents with business partner Franklin L. Pope.

Edison's shop on Ward Sreet, Newark, New Jersey, where he designed and manufactured telegraph inventions from 1871 to 1875.

telegraphs for the Gold and Stock Telegraph Co. By the fall of 1870, Edison was operating two manufacturing shops in Newark: the Newark Telegraph Works and the American Telegraph Works.

Newark was a thriving industrial city that manufactured, among other things, machine tools, jewelry, and chemicals. It also had iron foundries and, most important for Edison, a pool of skilled mechanics and machinists. Workers from Europe, particularly Germany, found ample employment opportunities in Newark's industrial economy. Most of Edison's Newark shop employees were expert German toolmakers, machinists, and mechanics. William Unger was of German descent, and machinist Sigmund Bergmann, who later worked closely with Edison in the electric light business, had immigrated to the United States from Germany as a young man. Charles Batchelor and John Kruesi were recent European immigrants who joined Edison in the early 1870s.

An English machinist who worked for the Manchester thread manufacturer J. P. Coates & Co., Batchelor arrived in the United States in 1870 to install machinery in a Newark thread mill. He remained in the United States after the assignment and began working for Edison's American Telegraph Works. John Kruesi was a Swiss-born machinist who spent his youth in an orphanage, where he learned weaving. He also served an apprenticeship with a locksmith before becoming a journeyman machinist in Zurich. Travel to Paris, London, Belgium, and Holland broadened his experience before he landed on U.S. soil in 1870 to work for the Singer Sewing Machine Co. in Elizabeth, New Jersey. Singer offered Kruesi advancement in the company after he improved its manufacturing operation, but the Swiss journeyman had heard about a promising young inventor in Newark named Edison and decided to work with him instead. "Kruesi," Batchelor said, "was the most indefatigable worker in the crowd. But he was also able to make things quickly and after to arrive at the most ingenious principles at once."

In early 1871 Edison's work on automatic telegraphy intensified as George Harrington brought additional investors into the American Telegraph Works. Meanwhile, the Newark Telegraph Works was busy producing printing telegraphs for the Gold and Stock Co. On May 26 Edison signed a lucrative contract with Gold and Stock that allowed him to relocate and expand his manufacturing shops. Under the terms of the agreement, Edison would work as

TOP: Charles Batchelor, Edison's principal assistant in the 1870s and 1880s. BOTTOM: Machinist John Kruesi, who built the first tinfoil phonograph in 1877.

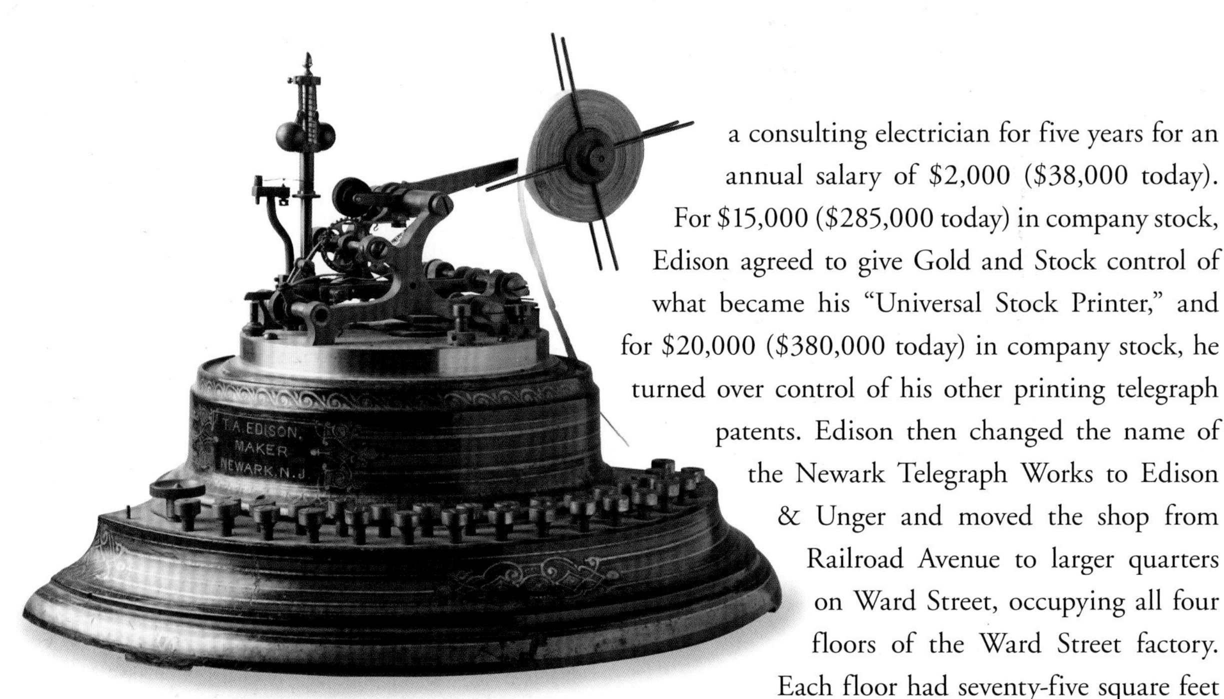

Edison's Universal Stock Printer transmitted stock and commodity prices to banks and brokerage houses.

a consulting electrician for five years for an annual salary of $2,000 ($38,000 today). For $15,000 ($285,000 today) in company stock, Edison agreed to give Gold and Stock control of what became his "Universal Stock Printer," and for $20,000 ($380,000 today) in company stock, he turned over control of his other printing telegraph patents. Edison then changed the name of the Newark Telegraph Works to Edison & Unger and moved the shop from Railroad Avenue to larger quarters on Ward Street, occupying all four floors of the Ward Street factory. Each floor had seventy-five square feet of work space.

Edison delivered to Gold and Stock a prototype of his Universal Stock Printer in June. Its design was similar to earlier stock printers, but Edison introduced several important improvements. The new model used less battery power and contained interchangeable parts, which made the assembly of stock tickers faster and more efficient. Edison also rearranged the printer's magnets, typewheels, printing arms, and other components so that operators could more easily adjust the instrument. Edison began manufacturing the printer by the end of the summer and, by the end of 1871, had produced 600 stock printers for Gold and Stock.

In October 1871 Edison established the News Reporting Telegraph Company, which offered subscribers news transmitted from a central office to printing telegraphs installed in homes. Anticipating twenty-four-hour news cycles by more than 100 years, the company promised customers "all general news of the world—financial, commercial, domestic and foreign—the moment such news is received in the main Telegraph Office in New York." The instruments, the circular noted, "are novel and very ornamental, are quite noiseless and require very little attention." The company would place the printer anywhere in the home without cost, but charged $3 a week for the service, ink, and paper.

LEFT TO RIGHT: Edison's first wife, Mary, and their three children, William Leslie, Marion Estelle, and Thomas Alva, Jr.

The News Reporting Telegraph Company failed within three months, but before it closed, according to an article published in a Pittsburgh newspaper in 1902, Edison noticed a female employee: sixteen-year-old Mary Stilwell. One day Edison approached Mary and asked her to marry him. "Don't be in a rush, though," he told her. "Think it over; talk to your mother about it and then let me know as soon as convenient. Say Tuesday." After talking to her mother, Mary accepted the proposal the next day, and they were married a week later on December 25, 1871.

(In an interview published in the June 1, 1884, edition of the *New York World*, Mary contradicted this story. She claimed that she never worked for Edison and that they first met in the spring of 1871 during a visit to his Newark shop. Edison courted Mary for several months before he proposed. Mary also refuted the view that Edison was a distant husband. "I have been very happy with him, and I expect to be as long as I live, for he is good and true and so tender to me and the children.")

Thomas and Mary had three children: Marion Estelle, born February 18, 1873; Thomas Alva, Jr., born January 10, 1876; and William Leslie, born October 26, 1878.

While there are no letters expressing their feelings for each other, notes made by Thomas in February 1872 suggest that he expected Mary to help him invent in the shop but was disappointed by her technical abilities. On February 1, in a notebook filled with sketches of telegraph circuits, Edison wrote, "Mrs. Mary Edison My Wife Dearly Beloved Cannot invent worth a Damn!!" Somewhat more playfully, on February 14 he jotted, "My Wife Popsy Wopsy Can't Invent."

WESTERN UNION PRESIDENT WILLIAM ORTON also influenced Edison in the early 1870s. Created in 1856, Western Union emerged from the Civil War as the largest telegraph company in the United States. It solidified its position in 1866 by merging with its two largest competitors—the American Telegraph Co. and the United States Telegraph Co.—but it faced competition from smaller regional companies and new competitors that arose following the expiration of the basic telegraph patents. Orton, like Marshall Lefferts, used new technologies to challenge his rivals. Unlike Lefferts, though, Orton was less a mentor to Edison than a savvy business leader who used Edison's technical abilities for the benefit of his company.

Edison may have met Orton as early as the spring of 1871, when a Gold and Stock executive gave him a letter of introduction to the Western Union president. At that time, Western Union had gained control of Gold and Stock because it wanted to move into the financial information business. The closer relationship between the two companies gave Edison better access to Western Union managers, who could provide him with technical support, access to information about the telegraph business, and funding to support his shops.

In the fall of 1872 Edison asked Western Union to support his experiments on duplex telegraphy, a system that allowed the transmission of two messages, in different directions, at the same time. Technologies that increased the capacity of existing wire networks without the need for the expensive construction of new lines appealed to telegraph companies like Western Union.

Edison met with Orton in February 1873 to discuss ideas for improving multiple telegraphy. The meeting resulted in an informal understanding that Edison would work for Western Union to develop multiple telegraph systems. In a December 1874 letter to duplex telegraph inventor Joseph Stearns, Orton explained why he supported Edison, mentioning his concern that his competitors would evade Stearns's poorly drafted patents:

> I employed him [Edison] to invent as many processes as possible for doing all or any part of the work covered by your patents. The object was to anticipate other inventors in new modes and also to patent as many combinations as possible. The object was not to glorify Edison, but to protect Western Union in the use of what they had purchased from you.

Designing duplex circuits was not a serious technical challenge for Edison. In early 1873 he invented several duplex circuits and designed diplex circuits for simultaneously transmit-

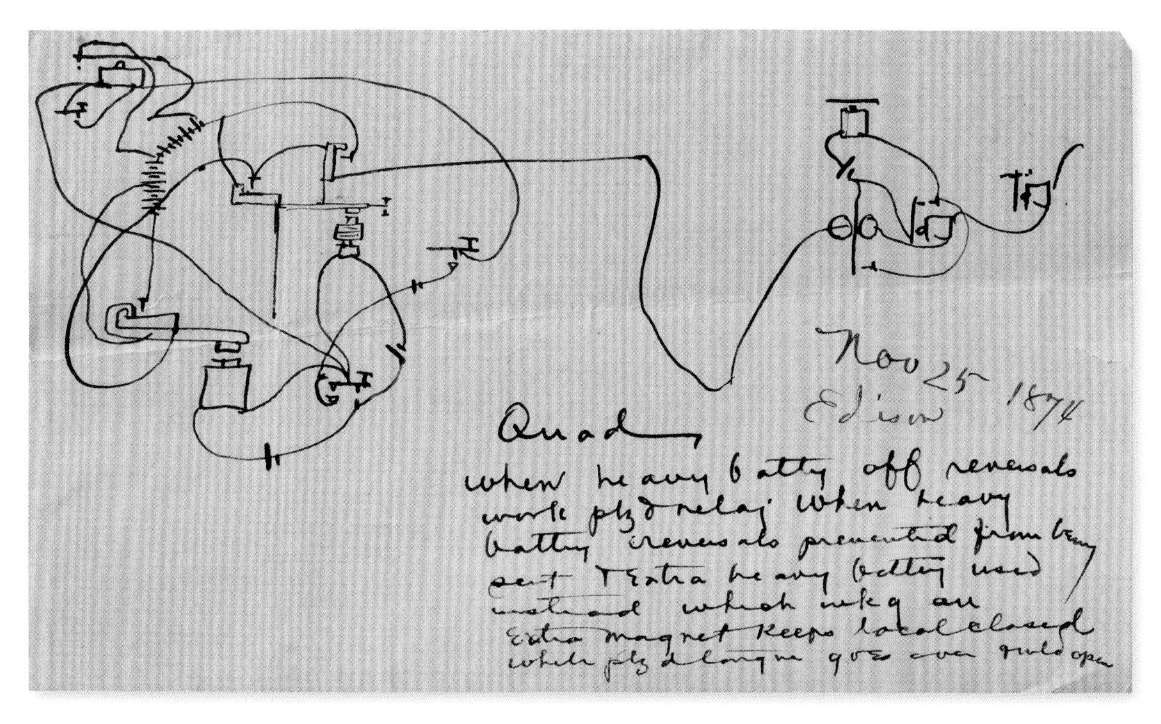

Edison's sketch of a quadruplex telegraph circuit, designed to send four messages along the same wire at the same time.

ting two messages in the same direction on the same wire. Edison also sketched ideas for what became the quadruplex telegraph, which combined duplex and diplex circuits to allow the simultaneous transmission of four messages on the same wire. Edison had accomplished Orton's objectives, but Orton did not formalize their relationship with a contract in 1873.

During a spring 1873 trip to England, where he tested an automatic telegraph system for the British Post Office, Edison devised a solution to a technical problem that prevented practical quadruplex circuits. In August 1873 he drafted a patent caveat, or preliminary patent application, for the quadruplex but, because he was busy with other telegraph inventions, did not pursue this research until the spring of 1874, when he asked Western Union chief electrician George B. Prescott for assistance in developing the system. With William Orton's encouragement, Prescott offered Edison the use of Western Union wires and facilities to test the quadruplex. News of the first test of the system between New York and Philadelphia appeared in the *New York Times* on July 10, 1874. By September, Edison had installed a quadruplex circuit on Western Union wires between New York and Boston and, in December, opened a circuit between New York and Chicago.

Financier and railroad promoter Jay Gould supported Edison's quadruplex experiments. Gould's purchase of the Atlantic & Pacific Telegraph Co. in 1874 was part of his effort to gain control of the American telegraph industry.

Edison needed money at the end of 1874. The lingering effects of an 1873 recession had caused telegraph companies to reduce equipment orders, and Edison needed cash to keep his shops operating and to pay his experimental expenses. Money was so tight that Edison had to move his family from their Newark home to a smaller apartment in the city's downtown section. He was hoping that the successful tests of the quadruplex would convince Orton to buy the technology and improve his finances.

Orton left New York at the end of December for an extended trip. Meanwhile, railroad financier Jay Gould had acquired a controlling interest in Western Union's rival, the Atlantic & Pacific Telegraph Co. As Orton left town, Gould acquired the Automatic Telegraph Co., including Edison's telegraph patents and his services as an electrical expert. On December 30, Gould visited Edison in Newark to discuss the quadruplex.

On January 4, 1875, Gould purchased Edison's quadruplex patents for $30,000 ($634,000 today). By this time, Edison had essentially completed the technical development of the quadruplex and had shifted the focus of his research toward automatic telegraphy. When Orton returned to New York later in January, he made Edison a counteroffer for the quadruplex patents. Edison refused to consider the proposal.

The money Edison received from Gould solved his immediate cash-flow problem, but it threatened his relationship with Western Union. At the end of January, Western Union filed a lawsuit against Edison, and soon Orton and Gould were involved in an intense legal conflict that involved the company's political allies at the highest levels of government in Washington.

More than anything else, Edison valued his autonomy. Amassing great wealth for its own sake had no appeal to him. Edison's ability to operate as an innovator was contingent on his access to money. With sufficient resources, he could accomplish a great deal. Without them, he could not pay the wages of his employees or purchase the tools, equipment, and supplies he needed to invent. As a result, he spent much of his time dealing with the problem of finding resources to keep his shops and laboratories in operation.

In the spring of 1875, Edison was successful enough to move toward greater autonomy. In May he ended the manufacturing partnership he had established with mechanic Joseph Murray in 1872 and opened a separate laboratory in the Ward Street factory. At the end of that month, Charles Batchelor drafted a list of proposed experiments that included, among other things, a new engraving process, a wireless electromagnet, a gold and silver detector, a cheap process for mining low-grade iron ore, and a new printing process. There was also "a copying press that will take 100 copies & system" on the list—a reference to the electric pen and duplicating press, Edison's first major nontelegraph-related invention. The new laboratory space and to-do list were signs that Edison was ready to move beyond the telegraph.

Despite his eagerness to bring new ideas to production, telegraph industry managers still needed Edison's services, and Edison needed their support. Orton's lawsuit over the quadruplex did not prevent him from meeting Edison in July 1875 to talk about acoustic telegraphy, a system that transmitted multiple messages by sending acoustic signals of different tones. Elisha Gray had invented an acoustic telegraph system, and Orton, concerned it would compete with Western Union's duplex systems, encouraged Edison to develop his own acoustic telegraph. Edison began acoustic research in the fall of 1875, and on December 14 Edison and Orton signed agreements that settled their dispute over the quadruplex and provided Western Union support for Edison's acoustic telegraph research. The agreement gave Edison the resources to open a new laboratory in Menlo Park, New Jersey.

"THE SUCCESSFUL PERSON MAKES A HABIT OF DOING WHAT THE FAILING PERSON DOESN'T LIKE TO DO."

THE ELECTRIC PEN

In 1875 Edison invented the electric pen, an instrument that allowed office workers to make stencils of handwritten documents. It was one of the first nontelegraph-related Edison inventions sold directly to consumers. The electric pen was a handheld reciprocating needle powered by a small electric motor and battery. The needle moved rapidly up and down as a writer wrote on a sheet of waxed paper, punching a series of small holes. The writer then made copies by passing an ink roller over the stencil mounted above blank sheets in a metal frame.

The electric pen reveals Edison's appreciation of public demand in the 1870s for improved office information technology. Insurance companies, railroads, and other large organizations needed efficient and affordable methods of reproducing multiple copies of forms, circulars, and other documents. Office workers could use carbon paper in typewriters to make up to ten copies, but before the electric pen, the only way to make multiple copies was a messy chemical stencil process on an apparatus called the papyrograph.

Edison and Charles Batchelor began copying experiments in the spring of 1875. At first, they attempted to make stencils using ink and chemically treated paper, but Edison was dissatisfied with the results. By the end of June, they "struck the idea of making a stencil of the paper by pricking it with a pen and then rubbing over with an ink." Edison and Batchelor designed a pen operated by a clockwork mechanism, but this also did not work well. By July, they had replaced the clockwork with an electric motor.

In September 1875, Edison signed a contract with his machinist, John Ott, to manufacture electric pens and began appointing sales agents in New York, Pennsylvania, and other states. Edison charged the agents $20 ($423 today) for each pen. The agents sold the pens, including batteries and accessories, for $30 ($634 today).

The electric pen was the first Edison invention sold directly to consumers.

Office workers were reluctant to accept the instrument and complained about the electric pen's noise and weight and the difficulty of maintaining its battery. As one agent noted, "I find that our greatest difficulty is to induce the clerks in places where we call to give it a trial. They do not like to take the trouble to learn to use the pen."

Consumer feedback helped Edison and Batchelor improve the electric pen's mechanical defects and design a battery that was easier to service. In November 1876 Edison assigned electric pen production and marketing rights to the Western Electric Manufacturing Co. of Chicago. Edison continued receiving electric pen royalties into the early 1880s, but by that decade the instrument faced competition from newer, less expensive, and more effective copying methods.

Edison's copying technology lived on in Albert B. Dick's stencil copier, the mimeograph. In 1887 Dick, a Chicago lumber merchant, purchased the rights to an 1880 Edison patent that covered a process for making stencils on a grooved metal plate covered with needle points. Dick's copying process used a modified typewriter to prepare stencils and a rotary press to make multiple copies. The mimeograph was commonly used in offices, schools, and churches to make inexpensive copies. The electric pen is also considered a forerunner of the modern tattoo needle. In 1891 New York tattoo artist Samuel O'Reilly altered the electric pen to inject ink into skin.

Edison's electric pen and duplicating press allowed office workers to make multiple copies of handwritten documents.

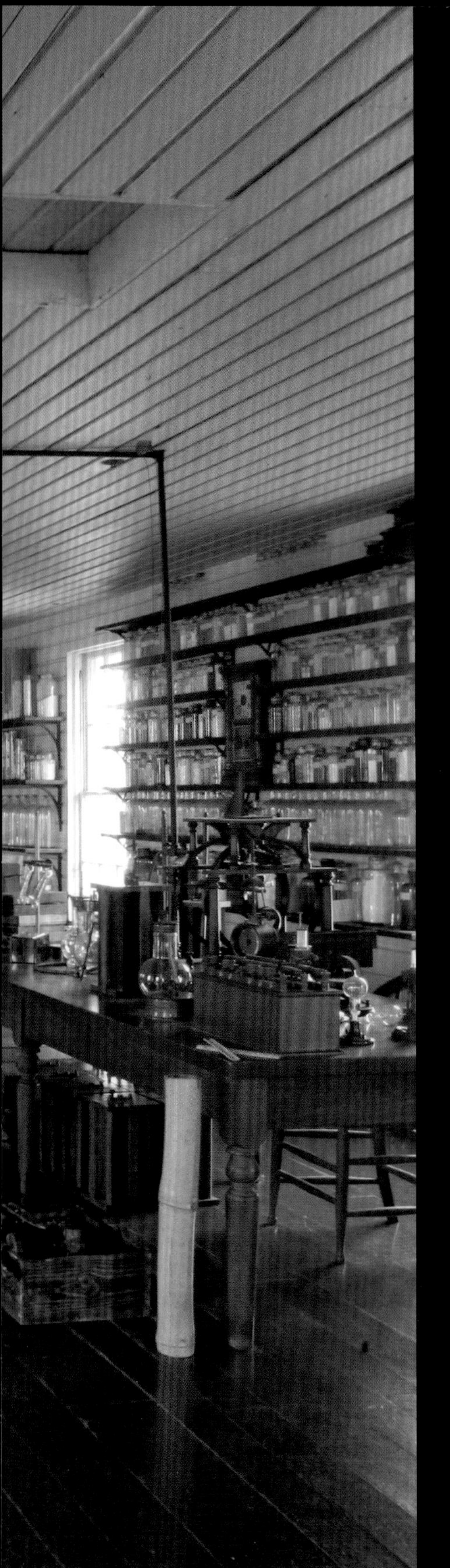

THE INVENTION FACTORY

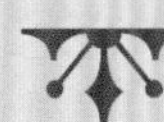

"HELL, THERE ARE NO RULES HERE—WE'RE TRYING TO ACCOMPLISH SOMETHING."

PAGES 18–19: Reconstruction of Edison's Menlo Park laboratory at Greenfield Village in Dearborn, Michigan. ABOVE: The Menlo Park lab (left) and the brick office and library (right) in 1880. The library housed the technical literature Edison gathered to support his electric light research.

THE MENLO PARK LABORATORY—where Edison and his experimenters improved the telephone, invented the phonograph, and developed a commercial electric light and power system—was the world's best-equipped private industrial research facility in the late 1870s and early 1880s. Edison's transformative inventions, by themselves, do not fully represent Menlo Park's significance, though. At Menlo Park, Edison combined three elements that shaped his approach to innovation for the rest of his life: team-based research, corporate support for research and development, and the branding of his persona as a reliable, practical inventor, in order to encourage investor financing of his laboratories and inventions.

A dispute with a landlord motivated Edison's move to Menlo Park. Edison had rented space in a Newark padlock factory on a monthly basis, and when he no longer needed it, he gave notice and returned the keys at the end of the month. The landlord sued Edison under an ordinance that made monthly renters responsible for a full year's rent. Outraged at the injustice of this law, Edison decided to leave Newark.

On December 29, 1875, Edison purchased two tracts of land in Menlo Park, a small village of 200 residents and a stop on the Pennsylvania Railroad, located twenty-five miles south of New York City and seven miles north of New Brunswick, New Jersey. On one of those tracts, between January and March 1876, Edison's father, Samuel, supervised the construction of a two-story laboratory. On Edison's other tract, closer to the railroad, stood a light-colored frame house with red trim, where Edison moved with Mary, their daughter, Marion, and their infant son, Thomas Jr.

In 1878 the *Philadelphia Times* described the Menlo Park laboratory:

> A frame tenement, nearly one hundred feet long, two stories high and without sign or ornament, it looks like a white frame country school house pulled out three times its own length. Two brick chimneys rise from one of its sides. A pond of rain water is just beside this building, and also one big tree, the whole enclosed by a paling fence, making a lot of an acre or more.

Edison's home at Menlo Park, summer 1881.

The first floor contained a drafting room and library, a machine shop, a blacksmith's shop, and a carpentry shop. *New York Sun* journalist Amos Cummings, who compared the laboratory to an "old fashioned Baptist tabernacle," colorfully described the second floor as

> an immense laboratory filled with electrical instruments. A thousand jars of chemicals were ranged against the walls. A circle of kerosene lamps was smoking on an empty brick forge. Their chimneys were the essence of blackness. An open rack loaded with jars of vitriol [metal sulfates] stood in the middle of the room, and the rays of the sun struck through them, flecking the floor with green patches. The western end of the apartment was occupied by telephones and other instruments, and there was a small organ in the southwestern corner.

Cummings spotted Edison at a table in the center of the room. "His hands were grimy with soot and oil, his straight dark hair stood nine ways for Sunday, his face was beardless but needed a shave, his black clothes were seedy, his shirt dirty and collarless and his shoes ridged with red Jersey mud."

Reporters asked Edison why he moved to Menlo Park. "I couldn't get peace and quiet and was run down by visitors," he answered. "I like it first rate out here in the green country and can study, work and think." Edison answered the question well enough but took the opportunity to express his antipathy toward urban life. "I may add that it is becoming the universal experience of men of professional or mental strain that they cannot stand the publicity of cities, the gassy hospitality and intrusive vealiness of adolescent club people, idiotic paragraphers and female men." Menlo Park's seclusion allowed Edison and his experimenters to avoid the distractions of cities. However, the laboratory's proximity to the railroad gave them easy access to the people and resources that made cities the centers of innovation.

Reporters nicknamed the Menlo Park laboratory the "invention factory," where Edison vowed to produce "a major invention every six months, a minor one every six weeks." But Menlo Park was not a factory. Factory owners valued productivity and discipline and used time clocks and rigid work rules to govern worker behavior on the factory floor. Worker autonomy and independence were discouraged.

Nineteenth-century machine shops, in contrast, were innovative environments. Machinists were skilled artisans who enjoyed more independence and autonomy than factory workers. They owned their tools and, in a system called "inside contracting," could negotiate compensation with shop owners. As a result, machinists had more control over the work process and production schedules than factory workers. Machine shops valued quality of workmanship over quantity and supported creativity by encouraging independence, informality, and individual initiative.

In December 1878, Edison asked Sarah Jordan, a stepsister of Mary Edison, to run a boardinghouse for his workers. Edison wired the house for twenty incandescent lamps in August 1880.

Edison and his staff on the second floor of the Menlo Park lab, 1880. Francis Jehl remembered that Edison often played the organ in a "pick and hunt style" during late-night suppers.

The informality of the machine shop work culture promoted new ideas and helped transfer technological knowledge. Following an apprenticeship under an experienced master, where they learned their trade, many machinists became itinerant, moving from city to city and shop to shop in search of work. As they moved around, they brought their skills with them. Independent inventors could rent access to tools and equipment and hire machinists to help them develop and test new inventions.

At Menlo Park, Edison brought together all the tools, equipment, information, and skilled artisans he needed to turn ideas into practical inventions. Although he relied heavily on the values and practices of the machine shop culture to utilize these resources in an innovative manner, Edison's ability to motivate experimenters to pursue a common agenda was key to his success. By providing leadership and determined laboratory goals, he impelled experimenters to pursue his ideas, not their own. As Edison explained in testimony for an electric light patent lawsuit in September 1881, he did not tell his workers how to solve problems; he expected them to think for themselves. "I generally instructed them on the general idea of what I wanted carried out, and when I came across an assistant who was in any way ingenious, I sometimes refused to help him out in his experiments, telling him to see if he could work it out himself, so as to encourage him."

Edison's principal Newark machinists, Charles Batchelor, John Kruesi, James Adams, and Charles N. Wurth, came with him to Menlo Park. Wurth was a Swiss machinist who had worked for the Singer Sewing Machine Co. before joining Edison in the fall of 1870. He had gone back to Switzerland in 1872 with a letter from Edison recommending him as "a steady and ingenious man," only to return to Edison in the fall of 1873.

These machinists formed a core group of about 200 employees who worked with Edison over the years between 1876 and 1882. During Edison's first two years at Menlo Park, there were no more than a dozen employees on the payroll. This number increased in 1878, when Edison hired more workers to assist with the electric light project. Workers came to Menlo Park for different reasons. Some wanted technical experience; others merely wanted jobs. Some of Edison's employees came only for a short time; others, like machinist John Ott, remained with Edison for the rest of their lives.

Work life at Menlo Park was not governed by the factory time clock. As Charles Clarke, an electric light experimenter, later recalled, "Laboratory life with Edison was a strenuous but joyous life for all, physically, mentally and emotionally. We worked long hours during the week, frequently to the limit of human endurance; and then we had time off from Saturday to late Sunday afternoon for rest and recreation."

The staff typically worked six days a week, ten hours a day, and sometimes late into the night. On those occasions Edison treated his workers to a midnight snack, followed by songs, jokes, and storytelling. Francis Jehl recalled, "During the midnight lunches, Edison often went to the organ and played a tune in the 'pick and hunt' style. One of the boys who could play would give us some of those old tunes of yesterday and all of us, including Edison, sang." These midnight breaks promoted a sense of community and camaraderie. Charles Clarke was impressed with the baskets of food brought in for the workers: "Not a cold lunch, mind you, but a hot dinner of flesh or fowl with vegetables, dessert and coffee. Hilarity increased

OPPOSITE: Edison (center, holding straw hat) with his experimenters on the porch of the Menlo Park lab, 1880.

EDISON'S NOTEBOOKS

Record keeping was essential to Edison's approach to innovation. It allowed him to document experimental work, preserve contractual rights with business partners, track the finances of his laboratories and companies, and secure patent protection for his inventions. Edison recognized the value of record keeping as early as 1870, when he wrote on the inside back cover of a pocket notebook, "All new inventions I will hereafter keep a full record."

Edison taking notes on rubber research at West Orange, December 1928.

Thomas Edison National Historical Park preserves more than five million documents relating to the development of Edison's inventions. At the heart of these records are the 3,500 standard notebooks Edison and his employees used to record laboratory experiments. In the early 1870s Edison kept notes on whatever was available, including ledger books, pocket notebooks, and loose scraps of paper. There was no attempt to systematize record keeping until the Menlo Park laboratory began electric light experiments in the fall of 1878. The complexities of designing the electric light system and the number of experimenters working on the project required Edison to adopt a standard six-inch-by-nine-inch laboratory notebook, which

Edison and his experimenters used to sketch ideas and track the results of experiments.

Edison often used laboratory records to protect and defend his patents against infringement, but the notebooks were more than just potential legal evidence; they enabled experimenters to communicate with one another and transfer ideas and concepts. Edison and his staff used rough sketches to explore solutions to technical problems on paper and to convey them to the machinists and mechanics in charge of creating three-dimensional models.

Because the standard-size notebooks were communal, many of them contain notes and sketches from multiple experimenters. For his own private notes, Edison used pocket notebooks as early as the late 1860s, when, as a telegrapher, he sketched ideas for improved telegraph circuits, tracked his spending on tools and supplies, and kept track of the books he read. There are 330 pocket notebooks in the Edison archives—most of them filled by Edison at West Orange from 1887 to 1931.

Edison's pocket notebooks allowed him to record ideas as he moved around the laboratory and between factories, conferring with experimenters and production managers. They contain page after page of to-do lists, instructions for his staff, ideas for new inventions, and experiments for improving existing products. They show that Edison did not innovate from one desk.

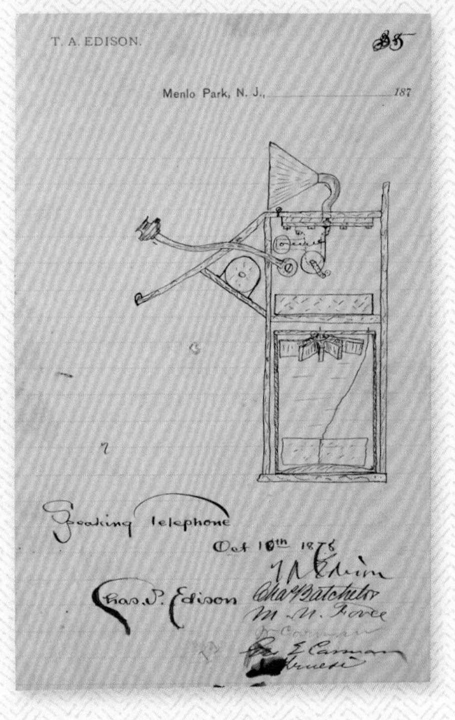
T. A. EDISON.

85

Menlo Park, N. J., 187

Speaking telephone

Oct 10th 1878

T A Edison

Chas P. Edison

Chas Batchelor

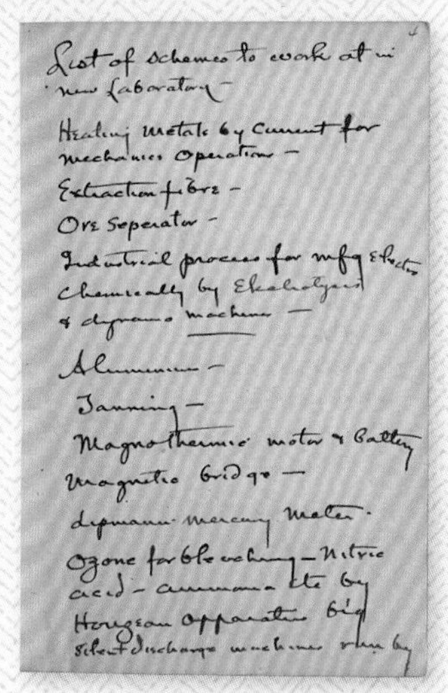
List of schemes to work at in new Laboratory —

Heating metals by current for mechanical operation —

Extraction ... ore —

Ore Seperator —

Industrial process for mfg Electro Chemically by Electrolysis & dynamo machines —

Aluminum —

Tanning —

Magno thermic motor & Battery

Magnetic bridge —

dynamo mercury meter.

Ozone for bleaching — Nitric acid — ammonia etc by ... apparatus big silent discharge machine ...

Edison's technical notes and drawings, like this telephone sketch and list of projects, allow us to see how his laboratories designed new inventions.

"I WORK NIGHTS IN ORDER TO ESCAPE FROM VISITORS. IT IS VERY NICE AND STILL HERE AT NIGHT!"

with the filling of stomachs, bantering and story telling were interlarded. At length Edison stood up, stretched, took a hitch at his waistband and began to saunter away—the signal that dinner was over, and work would be resumed."

Edison's capacity for hard work inspired loyalty and teamwork. "We work all night experimenting & sleep till noon in the day," Charles Batchelor wrote his brother in September 1875. "We have got 54 different things on the carpet & some we have been on for 4 or 5 years. Edison is an indefatigable worker & there is no kind of a failure however disastrous affects him." In April 1879 another experimenter, Francis Upton, wrote to his father about the difficulties of electric light research. "Mr. Edison will overcome them if any does. I have not in the least lost my faith in him for I see how wonderful the powers he has. He holds himself ready to make anything that he may be asked to make if it is not against any laws of nature. He says he will either have what he wants or prove it impossible."

Without corporate sponsorship, the Menlo Park laboratory would have been a well-equipped but isolated machine shop. Edison's track record as an inventor and manufacturer of telegraph technologies in the early 1870s demonstrated to the telegraph industry managers that invention was more than the ad hoc activity of unreliable inventors; it could be organized and managed to achieve the marketing goals of corporations.

William Orton, the president of Western Union, was a significant figure in this change. In the early 1870s independent inventors still developed most new technologies and, to market them, either sold their patents to existing companies or raised capital to organize their own companies. In 1870 Western Union created the position of electrician, which helped the company evaluate new telegraph technologies, but it did not have the ability to develop its own inventions. Edison's talents as an inventor, combined with his ability to create and manage productive and innovative laboratories, convinced Orton that supporting Edison would give Western Union a competitive advantage.

A contract Edison signed with Western Union in December 1875 to finance research on acoustic telegraphy gave him the resources to construct and equip the Menlo Park laboratory. On January 29, 1877, Edison drafted a letter to William Orton proposing regular financial

Edison branded his consumer products with his name, image, and distinctive umbrella signature.

support for the operation of the laboratory. "At present the cost of running my machine shop including coal, kerosene and labor is about 15 per day or 100 per week. At present I have no source of income which will warrant continuing my machine shop and I shall be compelled to close it unless I am able to provide funds," he explained to Orton. Edison proposed that Western Union pay him $100 per week ($2,220 today) to operate the machine shop in exchange for control of his telegraph inventions.

This proposal resulted in a five-year contract, signed on March 22, 1877, in which Western Union agreed to pay Edison $150 per week ($3,320 today) for laboratory expenses in exchange for control of his telegraph inventions. Western Union also agreed to pay Edison royalties on any inventions it adopted and to cover all patent and lawyer fees. In exchange, Edison was required to appoint Western Union's counsel, Grosvenor P. Lowrey, as his Patent Office representative. Lowrey later played an important role in securing the financing for Edison's electric light research (see Chapter 4).

Edison depended on his reputation as a reliable inventor to maintain investor confidence in the profitability of industrial research laboratories as a source of new technologies—and he relied on newspapers to promote this reputation. Americans in the late nineteenth century were fascinated by new technologies, and newspaper editors were eager to satisfy public curiosity by publishing articles about Edison's laboratories and inventions. Rather than cultivating publicity because of ego or self-aggrandizement, Edison gave reporters access to the laboratory and kept his name, image, and persona before the public for practical purposes. This early "branding" activity was directed not just to the public, but also to potential investors.

At times, Edison was uncomfortable with press attention, particularly when reporters interrupted important experiments. He also did not like sideshow-type publicity. This became

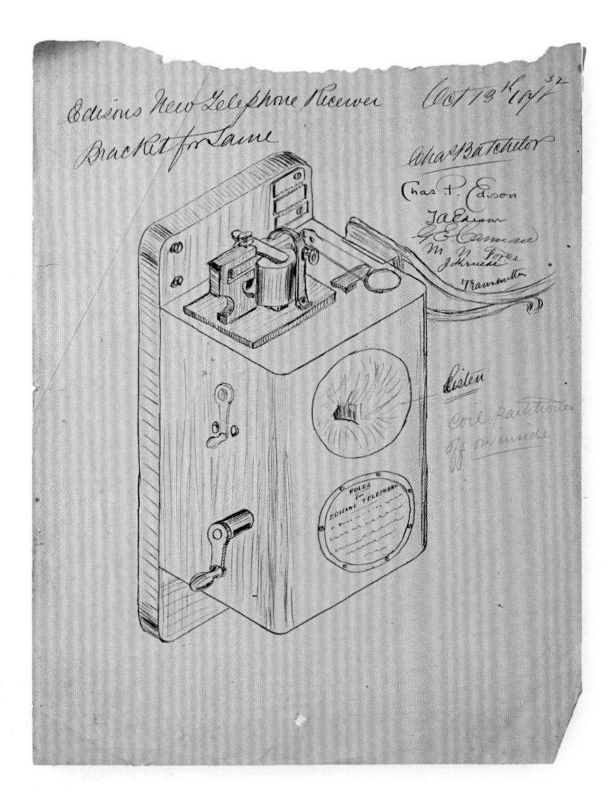

Carbon button telephone sketch. By carefully dating and signing his notes and drawings, Edison protected his ideas in case of patent litigation.

evident in March 1888, when he objected to the advertising schemes of his British phonograph agent, George Gouraud. "I am well aware of the value of the assistance which the press are capable of giving us, but it appears to me wholly unnecessary to make a parade after the fashion of Barnum and his white elephant."

Gouraud wanted to mount a phonograph exhibition in London and publish a book about Edison. Edison approved the exhibition but rejected the book idea. "I have no objection whatever to you advertising the phonograph to any extent you please, but personally I have no desire for notoriety and do not wish to be included in arrangements which are distasteful to me, and which, at least in America, would be undignified."

EDISON'S FIRST MAJOR EXPERIMENTAL PROJECT at Menlo Park was designing an alternative to Alexander Graham Bell's telephone. Edison's telephone research was based on his work with the acoustic telegraph, an instrument that used different sound tones to transmit multiple telegraph messages over one wire. Edison began acoustic telegraph research in the fall of 1875 with the financial support of Western Union. Edison and Elisha Gray—an electrical engineer employed by the Western Electric Manufacturing Co. in Chicago who also experimented with acoustic telegraphy—came close to developing an acoustic telegraph capable of transmitting speech, but their primary goal was the transmission of sound.

In June 1876, two Edison associates saw a demonstration of Bell's telephone at the Philadelphia Centennial Exhibition. Edison later claimed that he had begun working on the telephone in May (he executed a patent application for a "telephonic telegraph" on May 9), but reports of Bell's invention prompted him to intensify this work during the summer and fall of

1876. Western Union wanted to develop its own telephone network to compete with Bell, and by early 1877, the development of an alternative to Bell's instrument became a major research project at the Menlo Park Laboratory.

Edison's telephone work focused on improving Bell's transmitter. Bell's telephone used a metal diaphragm and a wire-wrapped magnet to transmit speech. The voice caused the diaphragm to vibrate, which produced an induced current in a wire, which re-created the vibrations at the receiving end and allowed a listener to hear the speech. Bell's telephone, however, produced a weak signal. Edison believed that this was caused by the weakness of Bell's induced current, and he experimented with ways to strengthen the signal by fluctuating or varying the current's resistance. To produce variable resistance, he developed a carbon-based transmitter.

Edison spent much of his time in 1877 and early 1878 searching for suitable forms of carbon and testing different telephone component arrangements. He also created more complex electric circuits than Bell used in his system, and also designed an induction coil, which allowed the telephone signal to be transmitted over longer distances.

In the spring of 1878, Western Union purchased the rights to Edison's telephone patents for $100,000 ($2.3 million today) and organized the American Speaking Telephone Co., which installed Edison's telephone in several U.S. cities. When Western Union sued the American Bell Telephone Co. for patent infringement, however, the courts upheld Bell's patent. In 1879, Western Union and the Bell Co. reached a compromise: in exchange for royalties on its patents, Western Union agreed to withdraw from the telephone business. Under this arrangement, Edison's telephone patents were transferred to the Bell Co.

Edison's "chalk receiver" telephone. According to legend, Edison's preferred telephone greeting was "Hello." Bell's choice, "Ahoy," never caught on.

THE TINFOIL PHONOGRAPH

"I'VE MADE SOME MACHINES; BUT THIS IS MY BABY, AND I EXPECT IT TO GROW UP TO BE A BIG FELLER AND SUPPORT ME IN MY OLD AGE."

POPULAR CULTURE OFTEN PORTRAYS the act of invention as a sudden flash of insight or inspiration that leads to new ways of doing things. Charles Batchelor described such a "eureka" moment when Edison discovered the idea of recording sound on July 17, 1877, while they worked on ways of storing and retrieving telephone messages. As he handled a telephone diaphragm—a thin membrane used to convert speech to electromagnetic waves—Edison proposed an experiment: "If we had a point on this," he told Batchelor, "we could make a record on some material which we could afterwards pull under the point, and it would give us speech back."

Batchelor attached a metal point to the center of the diaphragm and mounted it on a piece of grooved wood. Edison pulled a strip of waxed paper through the groove as he spoke

PAGES 32–33: Edison's original tinfoil phonograph, on display at Thomas Edison National Historical Park. The metal point, vibrated by a membrane inside the brass tube, indented sound waves on tinfoil wrapped around the cylinder. BELOW: Edison and his staff often tried many approaches until they found the right solution. In this March 1878 note, Edison drew different ideas for recording diaphragms.

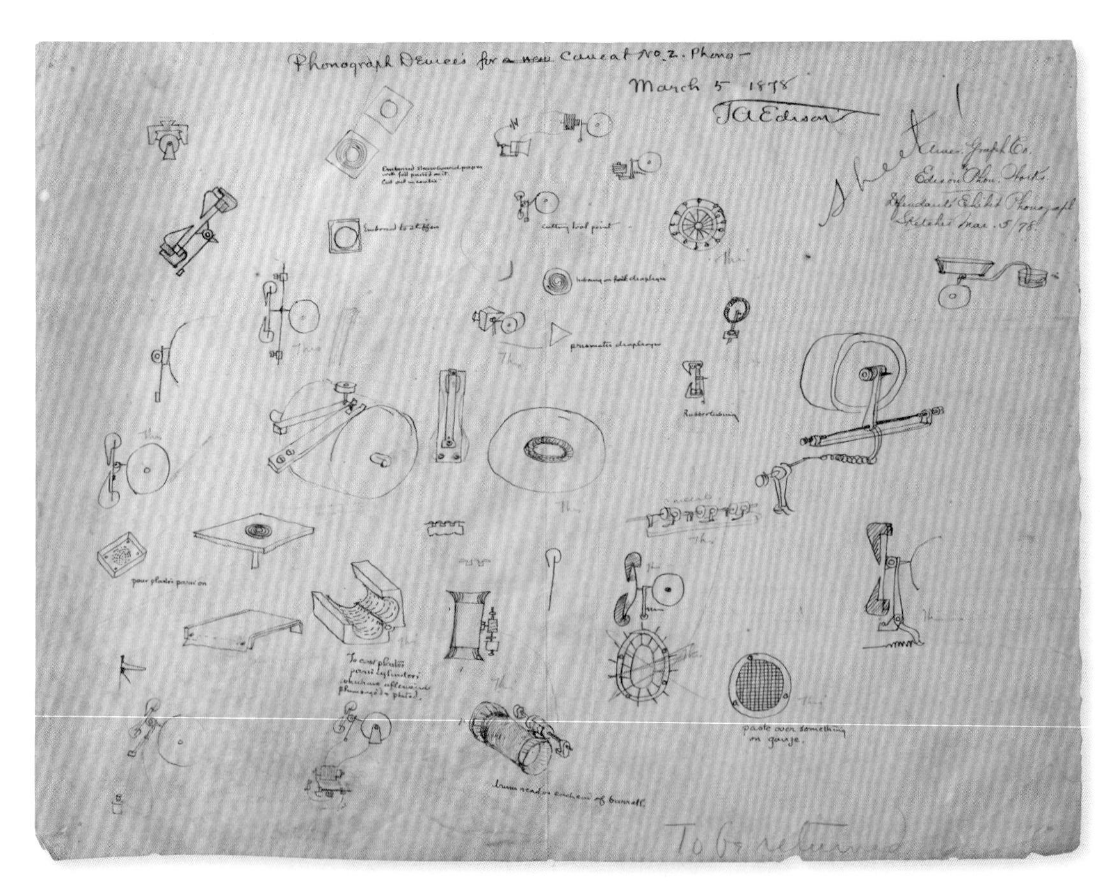

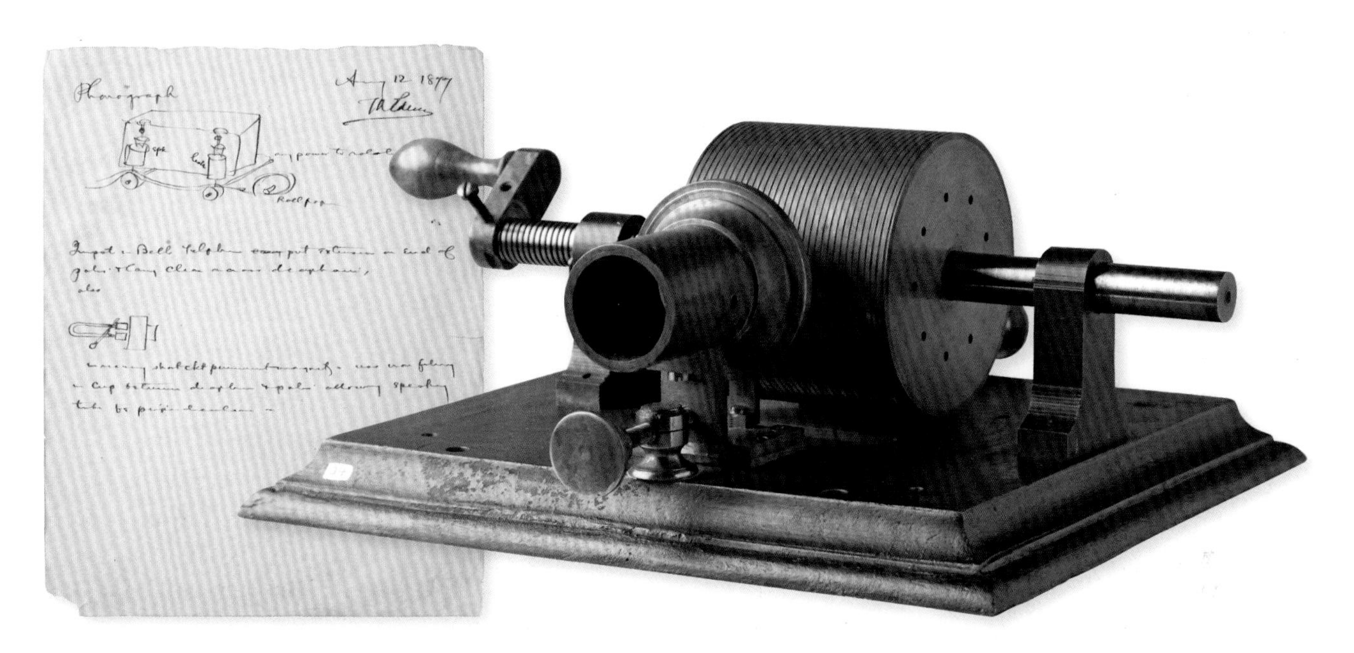

LEFT: Edison considered recording sound on strips of waxed paper in the summer of 1878 before eventually using sheets of tinfoil wrapped around a cylinder. RIGHT: The first tinfoil phonograph, now in the museum collection of Thomas Edison National Historical Park.

into the diaphragm. "On pulling the paper through a second time," Batchelor recalled, "we both of us recognized that we had recorded speech."

Edison's voice vibrated the diaphragm, causing the metal point to indent sound waves on the waxed paper. They reproduced speech by reversing the process. Edison noted the next day, "There's no doubt that I shall be able to store & reproduce automatically at any future time the human voice perfectly."

Edison's discovery of sound recording was a new insight shaped by prior experience—especially his work on the acoustic telegraph, the telephone, and the translating embosser, an instrument that recorded or "embossed" long telegraph messages on grooved paper discs placed on a revolving plate. Through his work on the acoustic telegraph and the telephone, Edison had gained familiarity with the physical properties and behavior of sound. Robert Spice, a Brooklyn chemistry teacher, had instructed Edison on acoustics in 1875, and Edison owned a copy of Herman von Helmholtz's influential study of acoustics, sound, and music theory, *On the Sensations of Tone* (1863). Edison may also have known of the phonautograph, an instrument invented in 1857 by French scientist Leon Scott that traced sound waves on a paper cylinder coated with lampblack (the sooty residue deposited on kerosene lamps).

Edison continued sound recording experiments during the summer and fall of 1878. He tried recording sound on paper of different thicknesses coated with various waxes and even attempted to record on the edge of the paper instead of the flat side. He also modified the size and shape of the recording points and diaphragms. In these early experiments, Edison tested different formats, including spools of paper tape and spirally grooved paper discs. By mid-September, he was focused on designing a cylinder recording machine, and by early November, he had envisioned recording on a sheet of tinfoil wrapped around a metal cylinder that would "indent about 200 spoken words & reproduce them from same cylinder." On November 29, Edison gave a sketch of this machine to John Kruesi, who spent six days making a model.

Kruesi finished the task on December 6. The tinfoil phonograph was a hand-cranked cylinder mounted between two diaphragms. Metal points attached to the phonograph with thin watch springs were placed between the diaphragms and the cylinder. On that day, Edison took a sheet of tinfoil, wrapped it around the brass cylinder, turned the crank, and spoke into the recording diaphragm. His first words were a verse from the children's nursery rhyme "Mary Had a Little Lamb." Edison's voice caused the diaphragm to vibrate the metal point, which indented sound waves on the tinfoil sheet. He reversed the process to listen to the recording, turning the cylinder back to the starting point, placing the reproducing point over the recorded groove, and turning the crank. To his astonishment, the phonograph worked on the first trial.

Edison and his staff wasted little time in publicizing the phonograph. Edison associate Edward Johnson described the process in a November 6 letter to *Scientific American*, the leading technical publication in the United States, and on December 7, Batchelor sent a description of the tinfoil phonograph to the editor of the *English Mechanic*. After Edison demonstrated the tinfoil phonograph for the editors of *Scientific American*, a report published in December by the journal described the event: "Mr. Thomas A. Edison recently came into this office, placed a little machine on our desk, turned a crank, and the machine inquired as to our health, asked how we liked the phonograph, informed us that *it* was very well, and bid us a cordial good night." The machine surprised the editors, who remarked, "No matter how familiar a person may be with modern machinery . . . it is impossible to listen to the mechanical speech without his experiencing the idea that his senses are deceiving him."

MARION EDISON
REMEMBERS MENLO PARK

Edison's oldest daughter, Marion, considered her childhood at Menlo Park one of the highlights of her life. As she recalled in 1956, shortly before her eighty-fourth birthday, "The laboratory had a queer fascination for me and I was never happier than when there." The lab's second floor and glassblower's house particularly fascinated her, but she never entered the machine shop, fearing that the machinery would catch her long blond hair.

"It was one of my daily duties to bring up father's lunch to the laboratory," she remembered. "I always peeked into the basket to see what he had and never failed to find a piece of pie so big that it covered the sandwiches."

Marion spoke most vividly about the phonograph. "I will never forget the day after the phonograph was invented. The men were all absent getting some rest after an exciting night. One lonely man was still there and when he turned the crank of the cylinder and 'Mary Had a Little Lamb' came clearly out in my father's voice, I jumped up and down for joy." She noted the streams of visitors who came to see the phonograph, including the famous French actress Sarah Bernhardt.

Of her home life, Marion remembered living in a three-story Victorian house run by three servants and a coachman "who lived in an apartment over the stable." When Thomas and Mary's third child, William Leslie, was born, on October 26, 1878, she was disappointed that the new baby was not a girl. "My brother Tom was a sickly, delicate child, and had to spend his winters in Florida, so was never a playmate."

Marion described her mother as a beautiful blonde who enjoyed giving lavish parties for her friends and family and wearing expensive clothes. "I still have photographs of some of the costumes Lord & Taylor made for her," Marion recalled. "One unusual one was a red and black brocade, decorated with stuffed red birds with black wings."

Despite the parties, Mary was lonely at Menlo Park. "My father neglected her for his work, or so it seemed to her. He never would come to her parties and he would often skip meals and very often would not come until early morning or not at all."

With Edison absent much of the time, Mary slept with a revolver under her pillow for protection. For the second time in his life, Edison was almost shot one night when he came home late. "Father forgot the front door key and not wanting to awaken the whole household, he climbed up the trellis onto the porch roof to his bedroom window. Mother, thinking he was a burglar, almost shot him. She let out a scream which father heard, then he called out to her, thus preventing a catastrophe."

Sarah Bernhardt asked to meet "le grand Edison" during her 1880 U.S. tour. During her December 5 visit to Menlo Park, an enthralled Bernhardt asked if Edison was married.

Following the *Scientific American* account, American and international newspapers published stories about the tinfoil phonograph and its inventor. Up until the end of 1877, Edison had been a well-known figure within the relatively small community of telegraph industry managers and engineers. While aware of Edison, the public probably did not know much about his work. The phonograph changed that. Press accounts created a sensation, bringing Edison international fame. As a result of this publicity, the Menlo Park laboratory was flooded with fan mail and requests for the new invention.

Several correspondents in early 1878 expressed interest in using the phonograph for scientific research. Stevens Institute of Technology professor Alfred M. Mayer wanted to use the phonograph in his acoustic experiments. As he wrote to Edison, "The results are far reaching (in science), its capabilities are immense & I cannot express my admiration of your genius better than by frankly saying that I would rather be the discoverer of your talking machine than to have made the best discovery of anyone who has worked in acoustics." Physician Clarence Blake thought the phonograph would be useful for studying speech and diseases of the larynx.

Alexander Graham Bell took an early interest in the tinfoil phonograph. He respected Edison's achievement but believed that his own work on the telephone had brought him close to discovering the principle of sound recording himself. As he wrote his father-in-law in March 1878, "It is a most astonishing thing to me that I should possibly have let this invention slip through my fingers when I consider how my thoughts have been directed to this subject for so many years past." Bell tested a phonograph to determine if it could be used for speech education but found the tinfoil recording inadequate for distinguishing certain vowel sounds, and he concluded that the phonograph was an interesting but deficient toy.

During the first months of 1878, Edison and his associates demonstrated the phonograph in Menlo Park, New York City, Washington, D.C., and other eastern U.S. cities. At the end of December 1877, Edison exhibited the machine for Western Union. Demonstrations were staged at Cooper Union and the Fifth Avenue Baptist Church in New York. Edward Johnson hosted exhibitions in Rhode Island and several upstate New York towns, including Elmira, Dunkirk, and Jamestown. "At all places great enthusiasm is shown over both the telephone

OPPOSITE: ***Scientific American*****, the leading technical journal in the United States in the nineteenth century, published the first description of Edison's tinfoil phonograph on December 22, 1877. News of the invention created a media sensation and made Edison a celebrity.**

system that will shorten the process of extracting the metals and reduce the cost, so as to enable poor ores, which are so abundant, to be worked at a profit. Millions of tons of the material are technically known as "tailings" (that is, ores from which has been taken all the gold and silver that, by present processes, can be profitably extracted, but which still contain an appreciable quantity of the precious metals) exist in all the auriferous districts. For the treatment of these ores various methods have been suggested. The principal difficulty that has been encountered is that of bringing mercury into contact with the gold where the latter exists in only small quantities, or from the flouring of the mercury when vapors of mercury are employed, entailing loss of amalgam and mercury in the subsequent treatment.

Messrs. Forster and Firmin, of Norristown, Pennsylvania, have recently devised a novel method of treating ores with mercury, for which letters patent have been granted them in the United States, Canada, Australia, and other countries. The pulverized ore containing free gold or silver is fed from the hopper, shown in the illustrations, with a horizontal tube, A, Fig. 2. While in the act of falling it is impinged

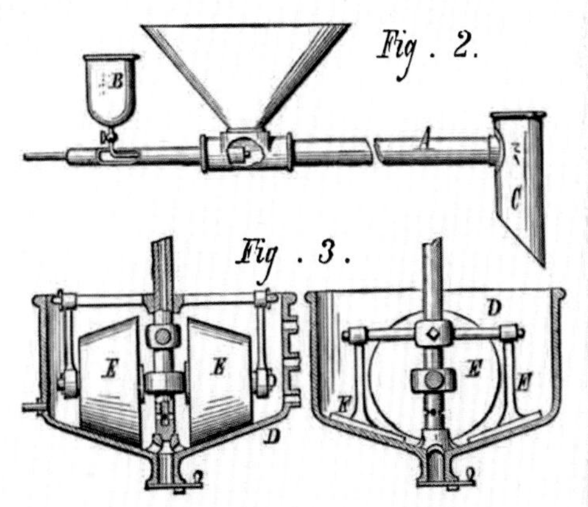

upon by a stream of mercury, which escapes from the receptacle, B, through the inner pipe shown. The flow is broken up and carried forward by steam or air pressure, after the manner of the well known principle of the sand blast. The horizontal tube connects with a vertical tube, C, upon which the ore and the atomized mercury are together forcibly projected, grain by grain, in a continuous stream, and fall, by their own gravity, into the washer or receiver, D. It is claimed that an almost unlimited quantity of ore may be treated by this process, as the attendants have only to feed the hoppers and remove the deposit. The inventors state that "with only a three inch tube from three to five tons of ore can be treated per hour."

In connection with this amalgamator an improved washer, shown in detail in Fig. 3, is used. This consists of a vessel, having a conical bottom, in which rollers, E, and also with scrapers or mullers, F, are placed. The feed water is injected through the shaft or near the bottom of the vessel, and the upward current carries off the waste ore, while the amalgam and surplus mercury collect in the dead water space in the conical bottom, whence they are drawn off through the discharge cock.

The advantages claimed for this invention are: 1st. The rapid continuous process of amalgamating, thus treating very large quantities of ore. 2d. The thorough impregnation of the metals with the mercury, giving larger results. 3d. The profitable working of poor ores or tailings, which are now valueless. 4th. The simplicity of the apparatus, having no parts to get out of repair. 5th. The cheapness and portability of the apparatus, and the ease and economy with which it can be operated wherever there is a steam boiler.

In the improved washer the amalgam and mercury are re-

Scientific American.

ESTABLISHED 1845.

MUNN & CO., Editors and Proprietors.

PUBLISHED WEEKLY AT
NO. 37 PARK ROW, NEW YORK.

O. D. MUNN. A. E. BEACH.

TERMS FOR THE SCIENTIFIC AMERICAN.

One copy, one year, postage included $3 20
One copy, six months, postage included 1 60

Clubs.—One extra copy of THE SCIENTIFIC AMERICAN will be supplied gratis for every club of five subscribers at $3.20 each; additional copies at same proportionate rate. Postage prepaid.

The Scientific American Supplement

is a distinct paper from the SCIENTIFIC AMERICAN. THE SUPPLEMENT is issued weekly; every number contains 16 octavo pages, with handsome cover, uniform in size with SCIENTIFIC AMERICAN. Terms of subscription for SUPPLEMENT, $5.00 a year, postage paid, to subscribers. Single copies 10 cents. Sold by all news dealers throughout the country.

Combined Rates. —The SCIENTIFIC AMERICAN and SUPPLEMENT will be sent for one year, postage free, on receipt of *seven dollars*. Both papers to one address or different addresses, as desired.

The safest way to remit is by draft, postal order, or registered letter.

Address MUNN & CO., 37 Park Row, N. Y.

☞ Subscriptions received and single copies of either paper sold by all the news agents.

Publishers' Notice to Mail Subscribers.

Mail subscribers will observe on the printed address of each paper the time for which they have prepaid. Before the time indicated expires, to insure a continuity of numbers, subscribers should remit for another year For the convenience of the mail clerks, they will please also state when their subscriptions expire.

New subscriptions will be entered from the time the order is received; but the back numbers of either the SCIENTIFIC AMERICAN or the SCIENTIFIC AMERICAN SUPPLEMENT will be sent from January when desired. In this case, the subscription will date from the commencement of the volume, and the latter will be complete for preservation or binding.

VOL. XXXVII., No. 25. [NEW SERIES.] *Thirty-second Year.*

NEW YORK, SATURDAY, DECEMBER 22, 1877.

Contents.

(Illustrated articles are marked with an asterisk.)

TABLE OF CONTENTS OF
THE SCIENTIFIC AMERICAN SUPPLEMENT
No. 108.
For the Week ending December 22, 1877.

Price 10 cents. To be had at this office and of all newsdealers.

THE TALKING PHONOGRAPH.

Mr. Thomas A. Edison recently came into this office, placed a little machine on our desk, turned a crank, and the machine inquired as to our health, asked how we liked the phonograph, informed us that *it* was very well, and bid us a cordial good night. These remarks were not only perfectly audible to ourselves, but to a dozen or more persons gathered around, and they were produced by the aid of no other mechanism than the simple little contrivance explained and illustrated below.

The principle on which the machine operates we recently explained quite fully in announcing the discovery. There is, first, a mouth piece, A, Fig. 1, across the inner orifice of which is a metal diaphragm, and to the center of this diaphragm is attached a point, also of metal. B is a brass cylinder supported on a shaft which is screw-threaded and turns in a nut for a bearing, so that when the cylinder is caused to revolve by the crank, C, it also has a horizontal travel in front of the mouthpiece, A. It will be clear that the point

Fig. 1.

on the metal diaphragm must, therefore, describe a spiral trace over the surface of the cylinder. On the latter is cut a spiral groove of like pitch to that on the shaft, and around the cylinder is attached a strip of tinfoil. When sounds are uttered in the mouthpiece, A, the diaphragm is caused to vibrate and the point thereon is caused to make contacts with the tinfoil at the portion where the latter crosses the spiral groove. Hence, the foil, not being there backed by the solid metal of the cylinder, becomes indented, and these indentations are necessarily an exact record of the sounds which produced them.

It might be said that at this point the machine has already become a complete phonograph or sound writer, but it yet remains to translate the remarks made. It should be remembered that the Marey and Rosapelly, the Scott, or the Barlow apparatus, which we recently described, proceed no further than this. Each has its own system of caligraphy, and after it has inscribed its peculiar sinuous lines it is still necessary to decipher them. Perhaps the best device of this kind ever contrived was the preparation of the human ear made by Dr. Clarence J. Blake, of Boston, for Professor Bell, the inventor of the telephone. This was simply the ear from an actual subject, suitably mounted and having attached to its drum a straw, which made traces on a blackened rotating cylinder. The difference in the traces of the sounds uttered in the ear was very clearly shown. Now there is no doubt that by practice, and the aid of a magnifier, it would be possible to read phonetically Mr. Edison's record of dots and dashes, but he saves us that trouble by literally making it read itself. The distinction is the same as if, instead of perusing a book

During an April 1878 visit to Washington, D.C., a proud Edison poses with his tinfoil phonograph at the Matthew Brady studio.

and phonograph," Johnson reported. "When they hear the phonograph reproducing my song with its imperfections they endanger the walls with the clamor."

In April Edison went to Washington, D.C., to present the phonograph to the annual meeting of the American Academy of Sciences, held at the Smithsonian Institution, and to members of Congress. He proudly posed with the phonograph in a photograph taken at the studio of Civil War photographer Matthew Brady. The Washington trip culminated with a late-night White House demonstration for President Rutherford B. Hayes. As Edison recalled for his authorized biography, "The exhibition continued till about 12:30 A.M., when Mrs. Hayes and several other ladies who had been made to get up and dress, appeared. I left at 2:30 A.M."

Edison enjoyed demonstrating the tinfoil phonograph, but questions about its commercial development required attention. At the end of December 1877, he appointed Theodore Puskas, a Hungarian promoter, as his European sales agent. On January 7, 1878, Edison signed two contracts covering the manufacture and marketing of phonographs for specialized applications. One agreement, with Oliver Russell, covered the use of toy phonographs. The other contract gave Daniel Somers and Henry Davies the right to sell talking clocks and watches. In exchange for these rights, Edison would receive a 10 percent royalty on each phonograph sold. Nothing came from these contracts.

On January 30, 1878, Edison gave a group of investors that included Alexander Graham Bell's father-in-law and Bell Telephone Co. founder Gardiner Hubbard the exclusive right to market phonographs in the United States. In exchange, Edison received a 20 percent royalty on each tinfoil phonograph sold and $10,000 ($233,000 today), which he agreed to apply toward developing the phonograph into a "satisfactory apparatus for dictating letters or reproducing musical compositions." On April 24, Hubbard and his partners organized the Edison Speaking Phonograph Co. in Connecticut.

This arrangement allowed Edison to outsource marketing responsibilities to company organizers and gave him the resources to improve the phonograph in the laboratory. Menlo Park constructed prototypes of different tinfoil phonograph models, but throughout 1878 Edison contracted out machine production to shops in Newark, New York, and Philadelphia. Shifting marketing and manufacturing to business associates and contractors enabled Edison to focus on technical design, an area of strength, but it also meant that he could not completely control sales strategy or production quality. Edison changed this approach in the early 1880s, when the complexities of introducing his electric lighting system required him to become more involved in manufacturing and marketing.

THE LACK OF a clearly defined application for the phonograph affected the company's ability to promote it. Edison's contract with the Edison Speaking Phonograph Co. mentioned entertainment and business applications but did not offer specifics. Edison and his associates considered a variety of applications for the phonograph. Along with using it as a telephone answering machine—as referred to in early experimental notes—Edison proposed several uses, including talking dolls and animals, music boxes, and talking clocks. Edison also speculated about employing the phonograph to record courtroom testimony and to play recorded sidewalk advertisements. In an article published in the May–June 1878 issue of *North American Review*, "The Phonograph and Its Future," Edison identified several potential applications, including speech education and talking books for the blind.

In April 1878 Edison told a reporter from the Philadelphia *Weekly Times* his idea of recording novels on the phonograph and described a "reading" experience similar to today's audiobooks. "Well, the phonograph will read poetry to you, if you are sick, or your eyes are weak or you cannot stand the light. It will read in dark or light. It will read a whole novel and re-read it so that ladies sewing can hear a story read off." Edison claimed that he could produce talking books at a much lower cost than printed versions. "I can furnish a novel pricked that way for three cents instead of fifty cents, the price of a paper novel, because the composition is not in type, but in punctures."

The tinfoil phonograph, however, could not do any of these things in 1878. Operating the instrument required a high level of skill and patience. The recording was not very loud or distinct, and the tinfoil sheet was delicate and easily damaged. Recordings could be reproduced only once or twice, and it was difficult, if not impossible, to store the recording for any length of time. Edward Johnson summarized the phonograph's condition in January 1878: "The phonograph is creating an immense stir but I think it impresses people more as a toy than as a practical machine."

Lacking a practical machine, the company decided to capitalize on the invention's novelty, but its managers debated whether they should sell or lease phonographs or establish a business demonstrating the phonograph to the public. Gardiner Hubbard preferred selling phonographs, but Edward H. Johnson, the company's general agent, pushed for exhibitions. Charles Cheever, another company manager, argued that "there could be quite a number of thousands of dollars made . . . by exhibiting the phonograph at once" and advocated licensing demonstrators in exclusive territories. Hubbard did not oppose phonograph exhibitions, but he believed that it would be difficult to organize and manage a national network of exhibitors. Responding to daily requests for machines and the need to generate revenue to pay mounting expenses, on April 2 Hubbard announced plans for agents of the American Bell Telephone Co. to sell large exhibition phonographs for $100 each ($2,330 today).

A few weeks later, Hubbard changed his mind about phonograph exhibitions. After attending a phonograph demonstration in Washington in mid-April, James Redpath, the organizer of the Redpath Lyceum Bureau in Boston, offered to establish a phonograph lecture business. Established in 1868, Redpath's Lyceum Bureau arranged for prominent authors and public figures—including Mark Twain, Henry Ward Beecher, and Frederick Douglass—to give lectures in cities and towns throughout the United States. It was a popular form of adult education and entertainment, and it was profitable for the bureau and the speakers. In the 1870s, new technologies, such as the telephone, were introduced to the public through traveling lectures and demonstrations. Hubbard, appreciating the popularity of public lectures, accepted Redpath's

"I HAVE NOT FAILED 10,000 TIMES. I'VE SUCCESSFULLY FOUND 10,000 WAYS THAT WILL NOT WORK."

offer of "introducing the phonograph to the public, by a series of lectures carried on simultaneously in different parts of the country."

Redpath began organizing phonograph exhibitions in May 1878. The Edison Speaking Phonograph Co. issued circulars, offering interested parties the exclusive right to exhibit the phonograph for up to three months in assigned territories. The exhibitors, who paid $100 for the privilege of demonstrating talking machines, charged the public twenty-five cents for admission. The company took 25 percent of the receipts.

Mark Twain was one of the most popular performers on the late-nineteenth-century lecture circuit. Before movies, radio, and TV, these lectures provided entertainment and introduced the public to new technologies, such as the telephone and phonograph.

In September 1878, Redpath reported that his exhibition business had earned over $7,000 ($163,000 today), but he closed his Irving Hall demonstration in New York City after two weeks because of poor attendance. "It has been liberally advertised but it does not pay," Redpath told Hubbard. "Mr. Johnson, of course, believes that if we hold on we shall soon have crowded houses. I don't. There have been upwards of 300 exhibitions of the phonograph in this city & I think the interest is exhausted."

George Bliss, who organized an exhibition business in Illinois, experienced similar problems. Bliss rented a store on Chicago's fashionable State Street to stage phonograph demonstrations and spent $200 ($4,650 today) to promote them through newspaper ads, street banners, and posters. However, he could not compete against more popular forms of entertainment and concluded that public interest in the phonograph had diminished.

By early January 1878 Edison had introduced several improvements to the tinfoil phonograph, including a larger cylinder, an amplifying funnel, and a flywheel to regulate the cylinder's speed. Further research in early 1878 focused on designing a small, inexpensive demonstration phonograph capable of recording up to fifty words. By March 1878 Edison was selling the smaller model for $30 ($698 today). He expected to sell a large number of these demonstration machines, but they did not work satisfactorily and only a few were sold.

The small demonstration phonograph was designed to generate revenue until the Menlo Park laboratory could complete what it hoped would become the standard model: a disc phonograph operated by a clockwork mechanism. Edison expected the clockwork phonograph to alleviate the problem of uneven speed produced by hand-driven machines. Steam-operated phonographs tested in January and February 1878 showed promise, but steam was not a practical power source. Edison completed a clockwork disc machine by early April, but it did not perform adequately, so he abandoned the disc format.

That month, Edison included a clockwork mechanism in a cylinder. Along with a clockwork motor, this machine featured a pendulum speed regulator, bearings to improve the positioning of the recording diaphragm and cylinder, and a moveable slide bar to attach the tinfoil more securely to the cylinder. On May 19, Edison wrote to Theodore Puskas, "the clockwork cylinder machine is running today & it works splendidly. I shall probably ship it next Saturday." Although Edison incorporated the clockwork design in a British patent specification, for unknown reasons he never marketed a clockwork cylinder phonograph.

Edison and Batchelor made their last serious attempt to improve the tinfoil phonograph at Menlo Park in early October 1878, when they sketched a design for a dictating machine featuring a vertically mounted cylinder containing a roll of tinfoil. By placing the foil inside the cylinder, Edison hoped to simplify the process of applying foil sheets to the machine. Sketches and patent drawings reveal that a clockwork motor would have powered the machine, which included controls to allow operators to start and stop recording at will. Edison told the *New York Sun* that this machine could record up to 4,000 words. "Any man can dictate to it at his leisure, and his office boy can run out a half dozen or dozen words, stop the machine by pulling the cord, and write out what is desired," he declared. The Edison Speaking Phonograph Co. hoped that this new version of the phonograph would revive their business.

Although the laboratory prepared measured drawings of this design and incorporated its features in a March 1878 patent application, Edison did not put the machine into commercial production.[1] By the fall of 1878 the Edison Speaking Phonograph Co. noticed a decline in sales and expressed concern about the public's disinterest in the phonograph. From October to November 1878, phonograph sales dropped from $2,305 to $1,065. In November Edison gave Edward Johnson a loan to gain control of the company from its original investors. Johnson planned to sell a small demonstration phonograph he was developing with Sigmund Bergmann, a former Edison employee who operated a manufacturing shop in New York, but interest in the tinfoil phonograph continued to diminish. Sales rose slightly in December but declined steadily between January and July of 1879. On January 21, 1879, the Edison Speaking Phonograph Co. canceled its contract with Edison. He received his last royalty statement in August 1879. By then, Edison was involved in electric light research. He would not return to the phonograph until 1886, shortly before opening his West Orange laboratory. The tinfoil phonograph failed in the market during the late 1870s for several reasons. The Edison Speaking Phonograph Co. did not focus on an application for the phonograph that would guide its technical improvement or enable the company to implement an effective marketing strategy. The company vacillated between different functions (entertainment vs. business) and different marketing policies (sales vs. leasing). In its efforts to improve the tinfoil phonograph, the Menlo Park laboratory focused on the machine, not the recording medium. Without a record that could be easily handled or stored—a problem Edison would address at West Orange—the phonograph was a machine of limited utility. Consequently, the Edison Speaking Phonograph Co. relied on limited and fleeting markets: a public interested in the phonograph as a novelty and a small group of scientists who looked to the talking machine as an instrument for acoustic research.

Amid public acclaim for the invention of the tinfoil phonograph, the *Daily Graphic* nicknamed Edison the "Wizard of Menlo Park" on April 10, 1878.

A BIG BONANZA: EDISON'S ELECTRIC LIGHTING SYSTEM

"THE ELECTRIC LIGHT IS THE LIGHT OF THE FUTURE—AND IT WILL BE MY LIGHT—UNLESS SOME OTHER FELLOW GETS UP A BETTER ONE."

IN THE FALL OF 1878, the Menlo Park laboratory began tackling a problem that inventors had unsuccessfully addressed since the 1840s: the development of a practical incandescent electric light. Edison and his team not only invented an incandescent lamp; they also designed a system for producing and distributing electric light and power and created companies to manufacture and market this system in the United States and other countries. These early Edison electric light companies were the basis of the modern electric power industry.

British scientist Humphry Davy discovered the principle of arc and incandescent lighting in the early 1800s when he found that certain materials, when heated to incandescence by electricity, emitted light. He also demonstrated that electricity flowing through a circuit connected to two carbon rods separated by a gap produced a bright light.

Russian engineer Paul Jablochkoff introduced the first practical arc lighting system in the late 1870s. The harsh arc lights, however, were more suitable for outdoor and large indoor

PAGES 46–47: Edison Mazda incandescent lightbulb in the West Orange lab stock room. General Electric, the successor to Edison's electric light manufacturing companies, began making bulbs under the Mazda trademark in 1909. ABOVE: In 1809 British chemist Humphry Davy invented the first electric arc light. Six years later he introduced a safe open-flame lamp for coal miners.

spaces. This limitation encouraged inventors to "subdivide" the electric light by designing smaller, less bright lamps for interior residential and commercial spaces.

Edison was aware of the problem. In September 1877 he brought a strip of carbonized paper to incandescence, but exposure to air quickly oxidized and burned the carbon. To prevent oxidation, Edison attempted to place the carbon in a vacuum, but he lacked an efficient vacuum pump and dropped the experiment.

On September 9, 1878, Edison traveled to Ansonia, Connecticut, with Charles Batchelor, University of Pennsylvania professor George Barker, and a newspaper reporter to visit William Wallace's brass factory. Earlier that summer, on a trip to the western United States, Edison and Barker had had long conversations about harnessing waterfalls to generate electricity and transmitting the power over long distances to operate mines. Barker had encouraged Edison to visit Ansonia to see Wallace's arc lighting system and new generator.

According to the *New York Mail*, Edison was excited by Wallace's system. He "ran from the instruments to the lights and from the lights back to the instrument. He sprawled all over a table with the simplicity of a child, and made all kinds of calculations." Edison returned to Menlo Park convinced that he could solve the problem of "subdividing" the electric light.

Edison's solution was a regulator that would prevent a lamp's element from melting or burning by momentarily cutting off the electric current before the element overheated. He envisioned electric lamps wired in a parallel circuit instead of in series, which would enable regulators to control their own lamps and allow consumers to turn off individual lamps without shutting down the entire system.

On September 13, Edison included these ideas in a patent caveat and telegraphed to Wallace: "I have struck a big bonanza." Batchelor wrote to a coworker, "We have struck a big thing on Electric Light & I think we have solved the problem of the subdivision of it so that we can make as many lights of small power as we like." Edison told the *New York Sun*, "I have it now. When it is known how I have accomplished my object, everybody will wonder why they have never thought of it, it is so simple."

Edison predicted that his electric light would provide illumination at a lower cost than gas lighting. His system would supply businesses and residences with light and power generated at a central location. He envisioned lighting large sections of lower Manhattan. "The same wires that bring the light to you will also bring power and heat. With the power you can run

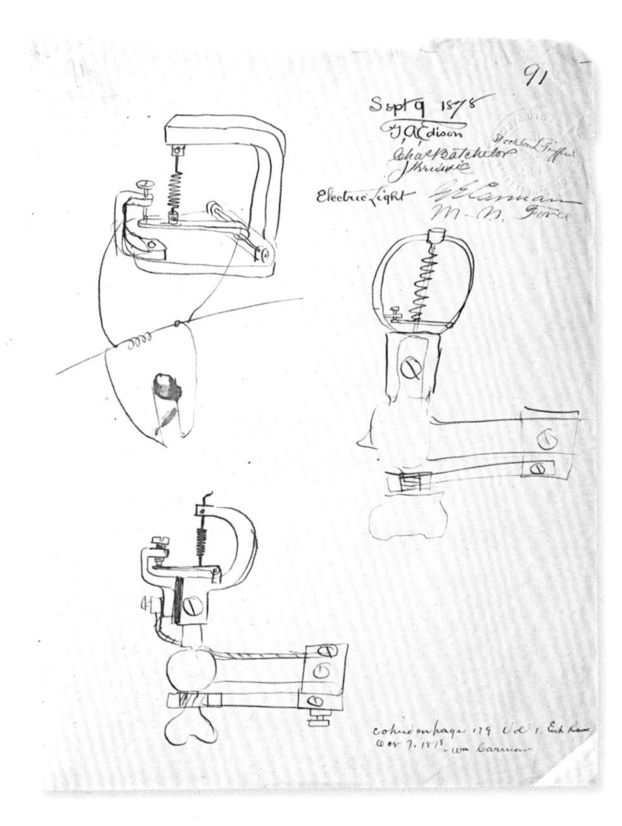

Edison drew these electric lamp sketches in September 1878, several months before he decided to put the burner inside a glass vacuum.

an elevator, a sewing machine or any other mechanical contrivance that requires a motor and by means of the heat you may cook your food," Edison told the *New York Sun*.

Following news reports of Edison's "breakthrough," the price of gas lighting company stocks declined on the New York and London stock exchanges. In late September, Edison's attorney, Grosvenor P. Lowrey, began negotiating with a group of potential investors that included Western Union president Norvin Green; Drexel, Morgan & Co. partner Egisto P. Fabbri; and business leaders associated with New York Central Railroad president William H. Vanderbilt. These talks resulted in the October 16 incorporation of the Edison Electric Light Co., capitalized at $300,000 ($6,440,000 today). Edison gave the company exclusive control of his North and South American electric light patents. In return, he received $250,000 in company stock, $30,000 ($698,000 today) for experimental costs, an additional $100,000, and a five-cent royalty on each lamp the company sold, with a guaranteed annual minimum of $15,000.

Lowrey—who was instrumental in creating the corporate organization to fund Edison's research—also acted as an intermediary between Edison and his investors. The directors of the Light Co., believing Edison had already solved the electric light problem, grew impatient when he failed to unveil his new lamp. Lowrey reassured Edison, who bristled at the lack of faith, and advised him to be honest with the company about the challenges he faced and arranged visits to Menlo Park for company officials.

Despite initial optimism, Edison was months away from introducing a practical electric light. As the Menlo Park staff worked on lamp regulators in the fall of 1878, they realized they would have to design an entire system—not just a lamp—including switches, meters, a distribution system, and, because Edison was dissatisfied with Wallace's dynamo, a new generator.

Edison also had to determine the system's technical requirements. This involved studying existing electric light technology and understanding the costs of gas and arc lighting systems. Edison hired Francis Upton to review the scientific and technical literature and the patents covering the field. Upton came to Menlo Park with an academic background in mathematics and science. He had degrees from Bowdoin College and Princeton and had spent a year studying in Berlin with German scientist Hermann von Helmholtz.

In November 1878, the laboratory designed its first light meter, which would allow electric light companies to measure customer electricity consumption. In December, experimenters began designing a new generator. Edison had identified platinum as the most promising material for the lamp's burner, but it was scarce and expensive. In January 1879 he began testing gold, iridium, nickel, and other metals to find a substitute, and concluded that platinum was still the best metal. He thus began a more systematic study of the properties of platinum—part of a larger effort to understand how platinum behaved under incandescent conditions.

In early 1879, Edison made two decisions that changed the direction of lamp research. After failing to prevent a platinum wire from disintegrating under incandescence, he decided to place the wire inside a vacuum. Edison had tried this in 1877, but he now had access to equipment that allowed him to produce better vacuums.

Headquarters of the Edison Electric Light Co. at 65 Fifth Avenue, New York City.

The other shift was Edison's decision to design a lamp with a filament of high resistance. (Resistance refers to the ability of a material, like copper or aluminum, to allow the passage of electricity.) Because of Ohm's law, elements in low-resistance lamps required lots of current to reach incandescence, which meant that conductors would have to be either very short or thick. A high-resistance lamp needed less current, enabling Edison to design longer, thinner conductors and reduce the overall cost of the system.

Placing lamp filaments in a high-vacuum glass globe prevented oxidation, but the expansion of gases in the platinum cracked the wire as the electricity heated it. Edison solved this problem by driving the gases off slowly as he gradually heated the wire in a vacuum. A greater challenge was finding an insulating material that would adhere to the platinum wire and prevent it from overheating.

By March, the laboratory had designed a lamp consisting of a platinum wire enclosed in a glass vacuum. Edison thought that he had succeeded in inventing a practical lamp and, after a small demonstration at Menlo Park late that month, began making plans for an even larger demonstration. His experimenters, however, had not yet found a way to prevent the platinum wire from melting. Another problem was the scarcity of platinum. In the spring of 1879, Edison sent hundreds of circular letters to postmasters in mining districts, asking for information about deposits of platinum ore. Work on designing generators, meters, and a distribution system continued during the spring and summer.

Despite progress on other system components, the laboratory had not yet designed a practical platinum lamp by the fall of 1879. The failure to produce an insulating material for the platinum wire was a significant obstacle. In early October the laboratory began experiments on filaments made out of carbon. Why Edison and his team tried carbon is unclear. It was a common material in nineteenth-century laboratories, and Edison used it in his carbon button telephone transmitter. In early October, Batchelor molded a spiral filament out of tar and lampblack and slowly baked it in an oven to carbonize it. Tests of the carbon revealed that it might meet Edison's requirements for high resistance. Later that month, Batchelor tested a variety of carbonized material, including fishing line, cardboard, and cotton soaked with tar. He achieved the best results with carbonized cotton thread. On October 22, 1879, a lamp with a carbonized cotton thread burned for thirteen and a half hours.

BATCHELOR'S LONG-BURNING CARBONIZED FILAMENT was the breakthrough Edison needed to complete a practical incandescent electric lamp, but it was not the end of research, as experimenters at Menlo Park continued to work on improving the lamp and other system components. On December 21, 1879, the *New York Herald* published the first account of the carbon lamp. During the last week of December, hundreds of visitors came to Menlo Park to see a light display that Edison had constructed at the laboratory. The railroad added extra trains to the schedule to accommodate the crowds. According to the *Herald*, "The laboratory was brilliantly illuminated with twenty-five electric lamps, the office and counting room with eight, and twenty others were distributed in the street leading to the depot and in some of the adjoining houses."

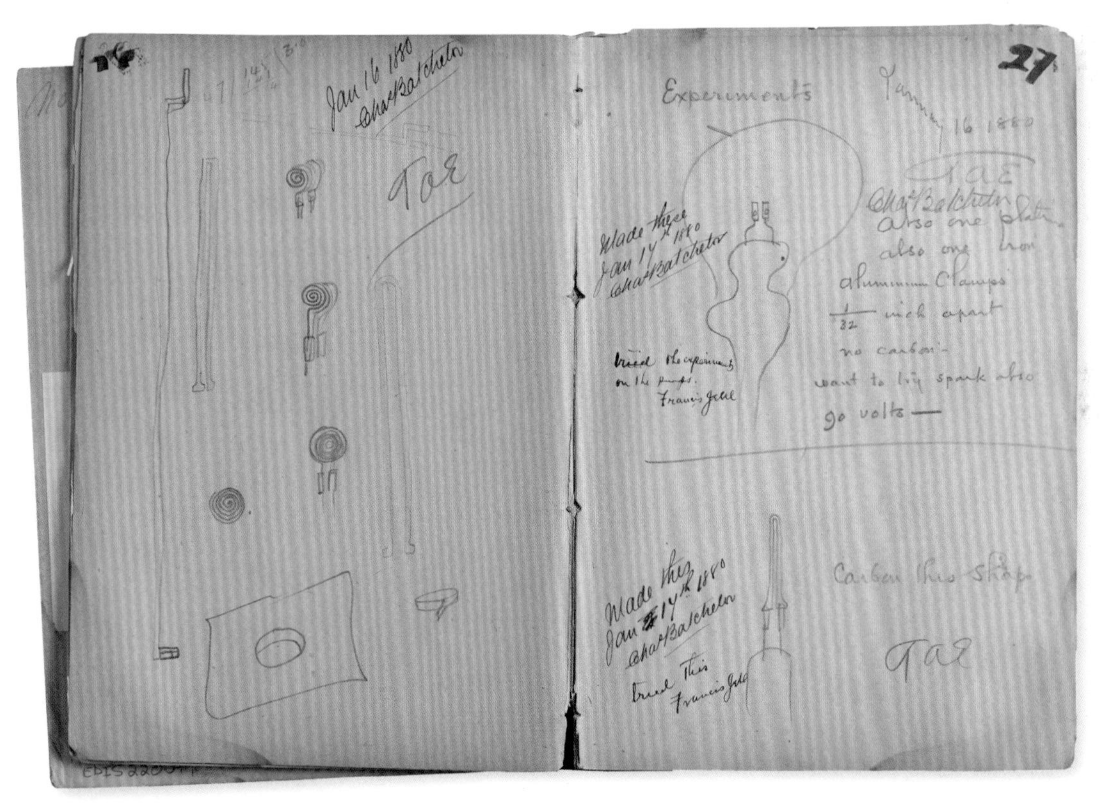

Edison's lab notebooks offer windows into the creative process. On these pages, Edison and Batchelor collaborate on the design of lamp filaments.

Early Edison lamps were simply plugged into a wooden base. The Menlo Park lab designed the more secure screw-type lamp socket familiar today.

From New York, news of Edison's accomplishment spread across the United States and around the world. Peter Dowd, a telegraph line contractor from Boston, wrote Edison on December 27: "Whenever I go downtown I am met by somebody or other who want to know what I think about your light . . . little of anything else, comparatively speaking, has been talked about in this city during the last week." Alfred Taylor wrote Edison from Chester, Pennsylvania, on January 3, 1880: "I have attentively read the numerous and various detailed accounts that have from time to time appeared in the newspapers, and am anxiously waiting the final step to be taken that will crown your indefatigable efforts with undoubted success to furnish to the world a cheaper and better light."

For Edison to furnish a better light to the world, he needed companies to manufacture and market the system. The Edison Electric Light Co. controlled Edison's patents, but its directors were reluctant to invest money in manufacturing operations, preferring instead to sell licensing rights to the new technology. Edison believed that he would have to manufacture the system in his own shops to control costs and to continue to improve electric lighting technology and production methods. Lowering production costs would help make the Edison electric lighting system more competitive.

In April 1880, Edison purchased an abandoned factory near the Menlo Park laboratory to manufacture lamps. Workers began producing vacuum pumps, while the laboratory staff continued improving the lamp. Menlo Park experimenters designed improved lamp sockets, developed techniques for testing defective filaments, and designed the tools to make filaments in large quantities. The laboratory also searched for a more reliable filament material. After testing a variety of substances, they discovered that bamboo fiber performed better as a filament than carbonized cotton threads. Edison sent agents to Japan, South America, and Florida to locate different bamboo varieties for testing. By December 1880, Edison had decided to use Japanese bamboo in his filaments.

The lamp factory, equipped with electric circular saws to cut bamboo filaments and gas-powered glass-blowing machines, produced its first batch of lamps for laboratory testing in September. In November, Edison organized the Edison Lamp Co. to manage the factory. With a capacity of 1,200 lamps per day, commercial production began in the spring of 1881.

On March 4, 1881, Edison organized the Electric Tube Co. to manufacture underground conductors for central stations. That same month Edison and Batchelor created the Edison Machine Works to produce dynamos and other large equipment for the Edison lighting system, and in April, Edison associates Sigmund Bergmann and Edward Johnson established Bergmann & Co. to produce electric light fixtures.

Edison had identified centrally produced electricity as a key marketing strategy of his system. In the October 1880 issue of *North American Review*, he announced plans for

Employees of the Edison Lamp Works at Menlo Park, New Jersey, 1880.

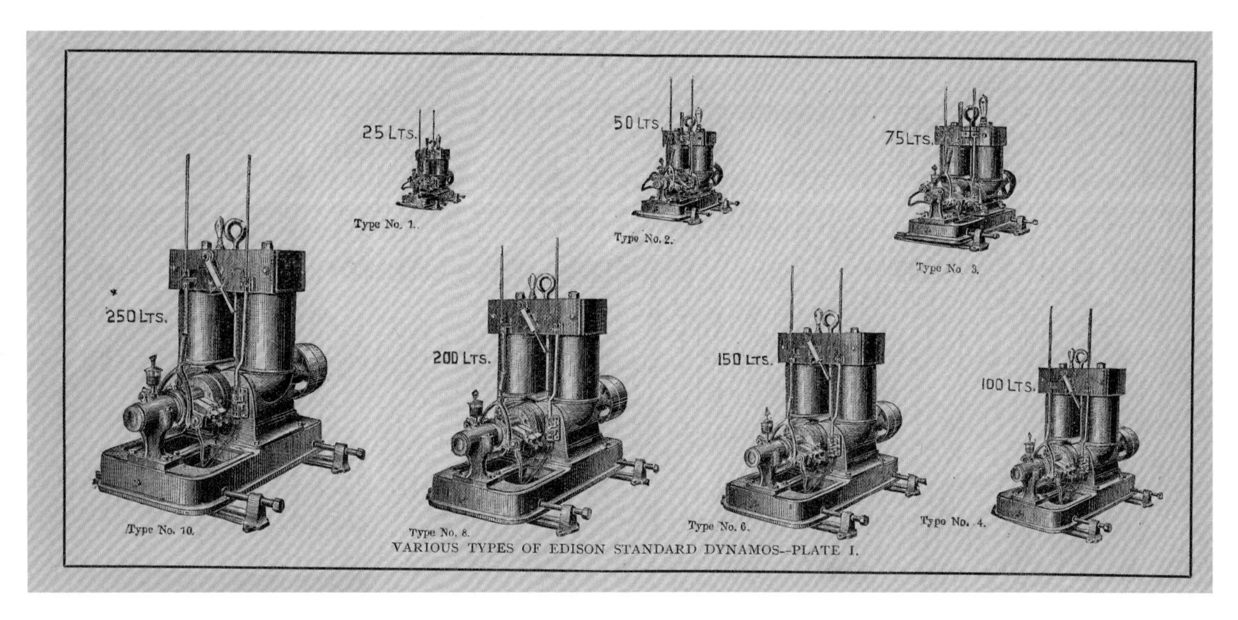

Edison dynamos, or generators, rated for different light capacities. Purchasers of isolated lighting plants could choose the dynamo suited to their needs.

introducing his electric lighting system "in all the great centers of population throughout the U.S." He described this system for the *New York World* in November:

> From a central depot in each district I shall send out light and power for half a mile. The wire will be laid in pipes. . . . Every 20 feet the pipes will pass through boxes, where connections may be made with houses. The wires will run directly to a meter in each building supplied, and the electricity will be measured just as gas is measured now.

Edison preferred to focus on developing a central station business, but he also offered customers isolated lighting plants—stand-alone systems with their own steam engines and generators. In May 1880, Edison installed his first isolated plant on the S.S. *Columbia*, a newly constructed steamship owned by Henry Villard's Oregon and Railway Navigation Co. As the first system operated outside of Menlo Park, it included four generators and 120 lamps. New York lithographers Hinds, Ketcham & Co. installed the first land-based isolated lighting plant in February 1881. The firm printed color graphics and claimed that Edison's light was better for matching colors than any other form of artificial illumination. In April,

Edison "jumbo" dynamo, or generator, at the 1881 Paris Electrical Exhibition. The dynamo weighed thirty tons and generated enough electricity to light 700 lamps.

the Edison Electric Light Co. created the Bureau of Isolated Lighting to manage the isolated plant business. The bureau became an independent firm—called the Edison Co. for Isolated Lighting—in November 1881.

By the spring of 1883, Edison had sold 330 isolated systems, which provided more than 64,000 lamps. Isolated plants were expensive and required an experienced technician to operate and maintain, so typically plants were sold to large commercial and industrial businesses, hotels, and theaters.

The Edison Electric Illuminating Co. of New York was organized on December 17, 1880, to construct and operate central stations in Manhattan. The first district was a one-mile-square area in lower Manhattan, bounded by the East River, Wall Street, and Spruce, Ferry, and Nassau Streets. Edison strategically located this first district to attract the attention of New York's newspapers and leading financial institutions, generating favorable publicity and the support of bankers who would invest in further development of the electric lighting system.

ELECTRIC RAILROAD

How do you keep electric power plants productive during the day, when no one needs electric lights? Edison answered this question at Menlo Park by designing motors to power elevators, industrial machinery, and electric railroads.

Edison first thought of electric railroads in the summer of 1878, as he traveled through Iowa on the way to California. Seeing Iowa's vast cornfields, Edison believed that short electric railroads could ship grain to the state's main railroad lines "and thus extend the radius of economic grain production."

In the winter of 1879, Edison asked Grosvenor Lowrey if the Edison Electric Light Co. would invest in an experimental railroad at Menlo Park. Lowrey told Edison to forget the idea and focus on the electric light, but Edison continued to work on the problem and, by May 1879, had produced a set of drawings for an electric locomotive and track. Edison later explained, "I determined to construct the railway the first chance I could get the money to do so."

Edison found the money in February 1880 and spent $15,000 ($340,000 today) building a three-fourths-of-a-mile-long track along sharp curves and steep grades next to the Menlo Park laboratory. The locomotive motor—a modified dynamo—had a capacity of thirty-five horsepower and could reach a speed of forty-two miles per hour. Power was supplied to the rails by two steam-driven dynamos in the laboratory's engine room. Workers completed the railroad in May 1880 and carried lab workers and visitors back and forth throughout the summer.

Edison began planning a longer railroad in August 1881. On September 14, Henry Villard, an Edison Electric Light Co. director who had recently assumed control of the Northern Pacific Railroad, agreed to finance construction of a new two-and-a-half-mile track near the Menlo Park laboratory, equipped with three cars and two locomotives—one for passengers, the other for freight. In November Samuel Insull reported that Edison

> is building a passenger locomotive which will be fitted up in splendid style and which will have a maximum capacity of one hundred miles an hour. Whether it will ever run at this rate when finished will very much depend upon the courage

of the driver. I think it would be a very good speculation to insure the lives of the passengers the first time Mr. Edison determines to run at this speed.

The two-and-a-half-mile–long railroad was completed by April of 1882. In May, Edison reported that he had "one locomotive capable of pulling four cars each containing thirty passengers at 20 miles per hour . . . the whole thing works splendidly."

Henry Villard, who needed funds to complete construction of the Northern Pacific Railroad, withdrew his support of the electric railroad experiments in December 1881, and Edison continued the research at his own expense throughout 1882. The lack of capital and the demands of dealing with his other electric lighting businesses, however, prevented Edison from organizing a company to market the electric railroad. In April 1883, he assigned his electric railroad patents to the Electric Railway Co. of the United States, to be managed by rival electric traction inventor Stephen D. Field.

ABOVE: Charles Batchelor at the controls of the Edison electric locomotive. Edison's daughter Marion remembered, "I was always very happy when riding on his electric railway." OPPOSITE: John and Fred Ott test an electric railroad motor outside the heavy machine shop of the West Orange lab in the early 1890s.

Edison needed the city of New York's permission to lay underground conductors, but the city government initially rejected Edison's request to install 100,000 feet of underground wiring. Favorable action on Edison's application, however, followed an electric light demonstration at Menlo Park for the New York City Board of Aldermen and an elaborate dinner catered by Delmonico's restaurant on December 20, 1880. Edison began laying underground conductors in the spring of 1881. To supervise construction of the system, Edison spent much of 1881 in New York. In March he moved his residence to New York and opened an office for his electric light companies at 65 Fifth Avenue.

Work crews completed installing the underground conductors in the summer of 1882. For the central station, the company purchased a building at 255-257 Pearl Street. To handle the weight of all the equipment, the foundation, walls, and floors of the building were

Edison at the dedication of a plaque on the site of the Pearl Street central station, New York City, October 1919.

fortified. Two eight-foot-tall smokestacks were constructed, along with a steam-operated conveyor system for loading coal into and removing ash from the boilers. The station had four large boilers and six twenty-seven-ton generators, each capable of producing 100 kilowatts of electricity. The Edison Electric Light Co. estimated that the plant would consume five tons of coal and 11,500 gallons of water each day. Construction costs for the station, including real estate, generating equipment, fixtures, and underground wires, reached $300,000 ($6,810,000 today).

The Pearl Street station began supplying electricity to 368 buildings wired for 8,117 lamps on September 4, 1882. On the next day, the *New York Herald* reported, "In stores and business places throughout the lower quarter of the city there was a strange glow last night. The dim flicker of gas, often subdued and debilitated by grim and uncleanly globes, was supplanted by a steady glare, bright and mellow."

By April 1884 the Pearl Street station was servicing 500 buildings, wired for 15,000 lamps. As the Edison Electric Light Co. announced in its bulletin that month, "The demand for the light far exceeds the supply, and the station is now being enlarged." Because operating expenses were higher than its income, the station lost money in 1882 and 1883, but made its first profit in 1884. The station operated until a fire on the morning of January 2, 1890, destroyed all but one of its dynamos. A rebuilt station continued operating until 1895, when it was decommissioned.

EDISON DID NOT CONFINE his electric lighting business to the United States. As early as December 1878, he had assigned the rights to his British electric light patents to J. P. Morgan's banking firm, Drexel, Morgan & Co. During the early 1880s, several companies were organized to promote the electric light in Europe, South America, and Asia. The Edison Electric Light Co. of Havana was established in June 1881. Edison's Indian & Colonial Electric Light Co., organized on June 13, 1881, controlled Edison's electric lighting patents in Australia and other British colonies. Buenos Aires received an electric light plant in August 1882. In May 1883, the Argentine Edison Electric Light Co. was organized.

NEXT PAGE: The successful display of Edison's electric lighting system at the Crystal Palace Exhibition in 1882 attracted favorable attention from English nobility and leading scientists.

Europe, however, was the focus of Edison's international electric lighting business. Edison sent Charles Batchelor to Paris in July 1881 to supervise the installation of an electric light display for the Exposition Internationale de l'Électricité. Late-nineteenth-century industrial exhibitions were important for introducing new technologies, and for Europeans and Americans, attendance at an exhibition gave them their first glimpse of new inventions, like the telephone, phonograph, and electric light. Exhibitions also allowed potential customers to compare competing technologies. Batchelor's assistant, Otto Moses, described the Paris exposition's opening: "With the greatest of pleasure I chronicle the complete success of our illumination. Last night for the first time we ran the entire capacity and I assure I never saw a more beautiful sight." French prime minister Léon Gambetta and King Kalakaua of Hawaii, among other prominent dignitaries, visited the Edison display.

Batchelor remained in Paris after the exposition to organize companies to manufacture and promote the electric light in France. Edison and his business associates were concerned that the interest generated by the Paris Exposition would encourage European competitors to infringe his patents. In February 1882 three electric light companies were organized in France: the Société Industrielle et Commerciale to manufacture electric lamps, the Société Électrique Edison to promote central stations in France, and the Compagnie Continentale Edison to license Edison electric light companies throughout Europe. One of these companies, the Deutsche Edison Gesellschaft, organized in March of 1883, operated a central station in Berlin.

Edison sent Edward H. Johnson to London in the fall of 1881 to exhibit the electric light at the Crystal Palace Exhibition and to open a central station. Johnson demonstrated the lamp at a meeting of the Royal Society of Arts in December. According to Johnson, "The Edison light made a successful debut in London . . . we had a crowded house and the lights were very steady and uniform."

At the 1882 Crystal Palace Exhibition, Johnson lighted two large rooms—the Entertainment Court and the Concert Room—as well as the avenue leading from the Crystal Palace to the exhibition railroad station. The concert room had 235 lamps, with eighty lamps suspended from a large chandelier. The Prince of Wales attended a preview on January 18, 1882. Johnson amazed the prince and 150 other dignitaries by submerging a lamp underwater and smashing another lamp wrapped in a cloth in order to prove that a broken lamp posed no fire danger.

The Holborn Viaduct central station served a half-mile-long district from Holborn Circus and Newgate Street to the General Post Office, powering nearly 1,000 lamps on four circuits for street lighting, restaurants, hotels, and shops.

Johnson thought that Londoners, who "live and breathe (foul gas-polluted air) in the daytime made artificially nighttime by the London fog," would appreciate Edison's clear, bright light. Johnson located the central station on Holborn Viaduct, a bridge built in the 1860s in the city of London connecting Holborn with Newgate Street. It was an ideal location for the central station because it was close to newspaper offices on Fleet Street, the General Post Office, two railroad stations, two hotels, and a church.

A subway ran under the viaduct, and gas and water mains serving the buildings on the viaduct were located in the tunnels. This allowed Johnson's workers to install power lines without opening the streets, which required the permission of Parliament. As a result, Johnson could supply electricity to the buildings without moving even a shovelful of earth. The station—the world's first Edison electric light central station—opened on April 13, 1882, with a capacity of 2,200 lamps.

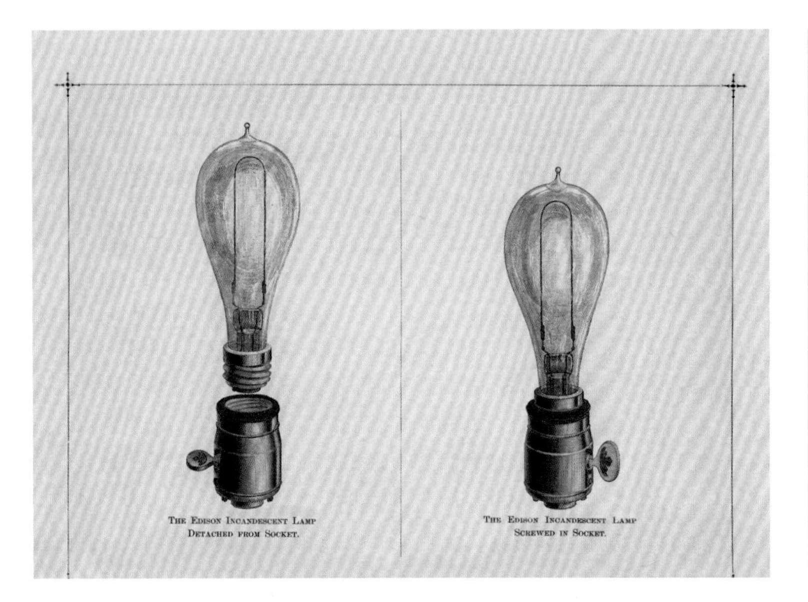

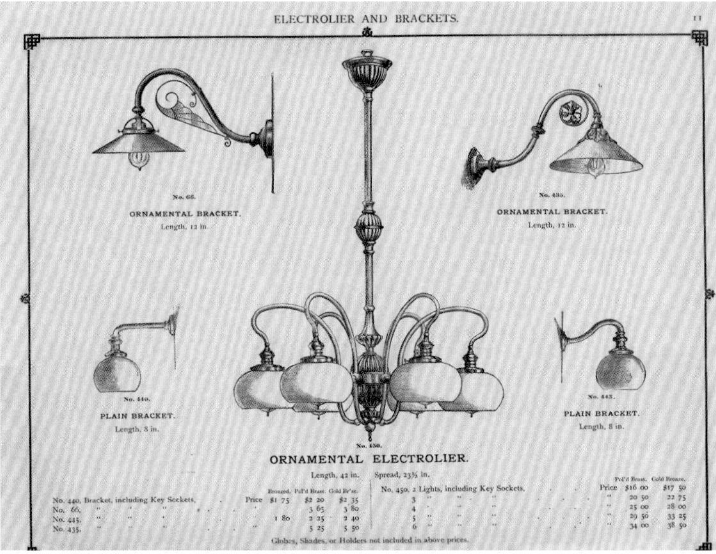

ELECTROLIER AND BRACKETS. 11

No. 66.
ORNAMENTAL BRACKET.
Length, 12 in.

No. 435.
ORNAMENTAL BRACKET.
Length, 12 in.

No. 440.
PLAIN BRACKET.
Length, 8 in.

No. 445.
PLAIN BRACKET.
Length, 8 in.

No. 450.
ORNAMENTAL ELECTROLIER.
Length, 42 in. Spread, 23½ in.

		Bronzed.	Pol'd Brass.	Gold Br'ze.
No. 440, Bracket, including Key Sockets,	Price	$1 75	$2 20	$2 35
No. 66, " " "	"		3 65	3 80
No. 445, " " "	"	1 80	2 25	2 40
No. 435, " " "	"		5 25	5 50

		Pol'd Brass.	Gold Bronze.
No. 450, 2 Lights, including Key Sockets,	Price	$16 00	$17 50
3 " " "	"	20 50	22 75
4 " " "	"	25 00	28 00
5 " " "	"	29 50	33 25
6 " " "	"	34 00	38 50

Globes, Shades, or Holders not included in above prices.

LEFT: Edison attempted to keep the cost of his electric light competitive with gas lighting, but the initial expense was high. Lamps were priced at $1 each ($22.70 today), but a 750-light isolated plant could cost as much as $12,000 ($272,000 today). RIGHT: Bergmann & Co. manufactured Edison electric light fixtures in different styles and finishes. The price of these fixtures ranged from $1.75 to $38.50 ($39.50 to $874 today).

By 1883 Edison had established a group of companies to manufacture and market his electric lighting system, but the parent company, the Edison Electric Light Co., lacked the capital to expand the business. On May 1, 1883, Edison invested his own money to organize the Thomas A. Edison Central Station Construction Dept. From May 1883 to September 1884, when it merged with the Edison Co. for Isolated Lighting, the Construction Dept. planned and built thirteen electric light central stations in smaller towns in Massachusetts, Ohio, Pennsylvania, and New York. In those towns, local investors organized an electric illuminating company and then paid Edison's Construction Dept. to design and install the electric lighting system. This involved canvassers conducting street-by-street, house-by-house surveys to collect detailed information about the number and types of customers. With this information, Edison's staff determined the best location for the town's central station and laid out the distribution system. Because Edison's direct current system could serve an area of only one square mile, the canvassers had to identify the right balance between residential and commercial customers in order for the local illuminating company to make a profit. The Edison Construction Dept. installed the generating plant and briefly operated the system before turning it over to the local company.

By early 1884 Edison had resolved the major technological challenges of designing an electric lighting system, but he continued to face a number of technical and commercial problems. In the laboratory he spent much of his time improving the performance of lamp filaments and lowering production costs. This was not pioneering research, but it was essential to the viability of his electric lighting business.

"I AM NOW A REGULAR CONTRACTOR FOR ELECTRIC LIGHT PLANTS, AND I'M GOING TO TAKE A LONG VACATION IN THE MATTER OF INVENTIONS."

In early February 1884 Thomas and Mary left New York for a two-month vacation in Florida. When they returned at the end of March, he confronted financial problems that threatened the expansion of his electric lighting business and eventually led to the reorganization of the Edison Electric Light Co. Edison had supported the Central Station Construction Dept. with $11,000 of his own money—funds he expected the Electric Light Co. to reimburse. The directors of the company, however, were never enthusiastic about being involved in the central station business, preferring to rely on the sale of patent rights to generate revenue. The company also lacked funds to repay Edison because of a shortage of capital in the nation's banking system, and the board of directors attempted to bring Edison's spending under control. Under these circumstances, Edison decided to close the Construction Dept. but instead reached an agreement with the Electric Light Co. to merge the department with the Edison Co. for Isolated Lighting on September 1, 1884.

In October Edison waged a proxy fight to replace the board of the Edison Electric Light Co. with directors who were more sympathetic to his business philosophy. As Edison told the *New York Sun,* "We want a Board with less law and more business. . . . I have given a perfect system, and I want to see it sold. . . . I don't want to see my work killed for want of proper pushing." In a compromise reached at the end of the month, Electric Light Co. president Sherburne Eaton and board member Grosvenor Lowrey resigned. They were replaced by Eugene Crowell, who became president, and Edison's former Menlo Park associate Edward Johnson, who became vice president.

5

FROM MENLO PARK TO WEST ORANGE

"TO INVENT, YOU NEED A GOOD IMAGINATION AND A PILE OF JUNK."

MARY EDISON DIED at Menlo Park on August 9, 1884, at the age of twenty-nine. For Edison, it was the first of a series of events in his personal life between 1884 and 1886 that shifted his world from Menlo Park and New York City to West Orange, New Jersey, and Fort Myers, Florida. When her mother died, Edison's daughter Marion remembered, "I found my father shaking with grief, weeping and sobbing." If Edison planned to remain in Menlo Park, as Marion later claimed, that changed with Mary's death.[1]

In the months after Mary's funeral, Edison traveled with Marion. Accompanied by his telegrapher friend Ezra Gilliland, they visited the New Orleans Cotton Centennial Exposition in February 1885. At the exposition, Edison met Mina Miller, the nineteen-year-old daughter of Ohio inventor, farm equipment manufacturer, and Chautauqua Institution cofounder Lewis Miller.

PAGES 68–69: Heavy machine shop at Edison's West to Orange laboratory, as it appears today. ABOVE: Edison winter home, Fort Myers, Florida, ca. 1909. Edison's quote to a 1914 local newspaper, "There is only one Fort Myers in the U.S. and there are 90 million people who are going to find it out," was adopted by Fort Myers as its motto.

In March, Edison, Marion, and Ezra Gilliland—joined by his wife, Lillian—traveled from New Orleans to Jacksonville, Florida. They stopped at St. Augustine before hiring a fishing sloop to take them down Florida's west coast to Punta Rassa, at the mouth of the Caloosahatchee River. From Punta Rassa they sailed up the river to the small town of Fort Myers.

As Edison later recalled, "The town was at that time mostly a cattle-town—mostly cattle and saloons, and the residents were mostly cattlemen and fishermen, cowboys were very common sights on the street." The locale and climate might have appealed to Edison. Fort Myers, the local newspaper boasted, "offers advantages unequaled at any point in Florida. Its mild, bracing atmosphere tones up the system and sends the glow of health to the cheeks. To the tourist the wild, varying scenery of the tropics, interspersed with visions of grazing cattle and traces of civilization, afford much enjoyment." Or maybe Edison appreciated the town's booster spirit. On the eve of his visit, the *Fort Myers Press* predicted a boom: "Jacksonville even now regards Fort Myers as a rival and other points are jealous of our superiority." In any case, before leaving Fort Myers he bought thirteen acres of riverfront property, about a mile from the town's business district.

Edison and Gilliland planned to build adjoining cottages and a laboratory on the property, allowing them to escape the harsh northeastern winters. In the fall of 1885 workers began clearing the land and constructing a dock and riverfront bulkhead. Lumber for the prefabricated houses was cut in Maine. The ship carrying the lumber from Boston to Florida stopped in New York to pick up furniture for the houses and boilers and machinery for Edison's laboratory. "The houses," the *Fort Myers Press* reported, "are two story buildings, square roofed, with a broad piazza around three sides, while a large kitchen is attached to both." They were completed in March 1886, during Edison's honeymoon with his new wife, Mina.

In March 1885, Edison and Marion returned to New York, and Ezra and Lillian Gilliland returned to their home in Boston. Edison visited Boston several times in the late spring and summer of 1885 to discuss an experimental contract with the Bell Telephone Co. and spend time at the Gillilands' summer cottage, Woodside Villa, on the north shore of Boston Bay. Mina Miller was a student at a Boston finishing school and a frequent visitor to Woodside Villa.

Mina and Thomas Edison, ca. 1908.

NO ONE KNOWS when Edison fell in love with Mina, but in a diary he kept in July 1885, he described being distracted by a vision of Mina on a Boston street. "Saw a lady who looked like Mina—got thinking about Mina and came near being run over by a street car—If Mina interferes much more will have to take out an accident policy."

During their courtship Edison taught Mina Morse code so that they could communicate in private. An August visit to Chautauqua in upstate New York, where the Miller family had a cottage, gave Edison another chance to see Mina. He invited Mina to join him, Marion, and the Gillilands for a trip through upstate New York and New Hampshire, and at the end of the summer Edison tapped out a marriage proposal in Morse code on Mina's hand during a carriage ride through the White Mountains. She tapped back, "Yes."

Thomas and Mina were married at the Miller family home in Akron, Ohio, on February 24, 1886. Edison family lore holds that Edison gave Mina a choice of residences: a fashionable town house in New York City or a home in the country. Mina opted for a country house. Shortly

Edison bought Glenmont, which included the twenty-nine-room Queen Anne–style house, a barn, a greenhouse, and thirteen acres of land, for $125,000 (about $3 million today).

before the wedding, Edison purchased Glenmont, one of the largest estates in Llewellyn Park, an exclusive residential community in West Orange, New Jersey. Designed by architect Henry Hudson Holly and constructed between 1880 and 1882, Glenmont was the home Henry C. Pedder, an executive for New York dry goods merchants Arnold, Constable & Co. In July 1884 Arnold, Constable accused Pedder of embezzlement and seized Glenmont. Eager to unload the house, on January 11, 1886, Arnold, Constable sold Glenmont—which included thirteen acres of land, a barn, a greenhouse, and several other outbuildings—to Edison for $125,000 ($3,080,000 today). Awed by the elaborate house and sprawling grounds, Edison said, "It is a great deal too nice for me, but it isn't half nice enough for my little wife here."

Madeleine Edison married John Sloane in Glenmont's first-floor drawing room in June 1914. The family held a private funeral service for Edison here on October 21, 1931.

At Glenmont Mina and Thomas raised their three children: Madeleine, born May 31, 1888; Charles, born August 3, 1890; and Theodore, born July 10, 1898.

Mina was in charge of Glenmont. She hired and supervised the household staff, planned meals, purchased household supplies, paid bills, and dealt with repairs and maintenance. Being the young wife of a busy inventor who worked long hours was a strain for Mina. She was often lonely, especially during the 1890s when Edison spent long stretches of time working at his ore milling plant in the hills of northwestern New Jersey. But Mina had the education and upbringing to run an active household and uphold the Edison family's social responsibilities. She defined her role as Glenmont's "home executive," and her goal was to provide Edison with a supportive home environment. As she told the magazine *Collier's* in 1925, "My job has been always to take care of Mr. Edison—to take care that his home contributed as much as possible to his doing the work that he had to do to the best advantage.

Thomas and Mina's three children, left to right: Theodore, Madeleine (with her nurse, Helena McCarthy Doyle), and Charles.

We have always put his work first, all of us. And we have tried to organize our home life to give results just as much as the laboratory."

In an oral history interview recorded several years before her death in 1979, Madeleine was asked what she thought of all the books that had been written about her father. She liked some of them, but none were satisfactory, in her view, because they neglected his family life. From reading these books, she remarked, "You would think he was just sort of a robot and never stopped working—well, that wasn't true."

In one of Madeleine's earliest memories of her father, she accompanied Mina to the Orange train station at the age of five to meet Edison upon his return from his ore milling plant in Ogden, New Jersey. "The most bedraggled man I have ever seen got off that train, and just covered in soot. I thought, my goodness, is that my father?"

Mina disciplined the children, but Thomas had definite ideas about how his children should be raised. One morning, he saw Madeleine lounging in a chair in her bedroom while a maid buttoned her shoes. Edison was furious. He told Mina, "No child of his would have her shoes buttoned by the maid, she'd button them herself."

According to Madeleine, Edison liked the company of children and young people, but he disliked formal entertaining. Invariably, before one of Mina's dinner parties, Edison would develop a severe case of indigestion. Mina would ask Madeleine to sit with the ill inventor upstairs, "to take his last words," while she hosted their guests.

Thomas and Charles Edison fishing at Bellaire, Florida, in 1900. In 1940, Charles wrote on the back of the photo, "We had been out in a small sailboat around the coral reefs. The biggest fish was mine!"

Edison spent Sunday mornings reading as many as five newspapers. In the afternoons he took walks with the children or enjoyed automobile rides—a favorite pastime. On Sunday evenings, the Edisons welcomed friends and members of Mina's family, who gathered around the piano to sing songs or act out scenes from Shakespeare. At Christmas, the house would be filled with up to thirty guests, and during the winter the family regularly attended Tuesday night vaudeville shows in Newark. Mina went with misgivings. Driving past the Methodist church on the way to Newark reminded her that Tuesday was prayer meeting night.

When Thomas and Mina moved into Glenmont in 1886, he used space in the Edison Lamp Works in Harrison, New Jersey, for research. Edison planned to continue working on the electric light, but he had other ideas for new inventions and soon decided that he needed a new, larger laboratory closer to Glenmont.

IN JANUARY 1887, Edison purchased two acres of land in West Orange at the corner of Valley Road (now Main Street) and Lakeside Avenue to construct his new laboratory. To design the new facility, he hired Glenmont's architect, Henry Hudson Holly, who submitted plans for a three-story brick building with 37,500 square feet of work space. Construction began in May.

Later that spring, during a site inspection, Edison discovered that workers had built the lab's walls one inch out of plumb. He promptly fired Holly and hired a new architect, Joseph Taft. During the summer, Edison decided that the building did not have enough research space, so he instructed Taft to add four one-story brick buildings—each 100 feet by twenty-five feet—to the original plan.

The West Orange laboratory was much larger than the Menlo Park lab, but it also reflected a significant shift in Edison's approach to innovation. Edison still owned a controlling interest in his electric light manufacturing companies—then based in Schenectady, New York—and he planned to continue electric light research at West Orange. He also envisioned, however, developing a variety of simpler technologies that could be easily manufactured and marketed. "My plan contemplates to working on only that class of inventions which require but small investments for each and of a highly profitable nature & also of that character that the articles are only sold to jobbers. No cumbersome inventions like the electric light."

Aerial photograph of Edison's West Orange laboratory and manufacturing complex in the late 1920s.

Diversifying the products of his laboratory allowed Edison to apply ideas and concepts from one invention to another, which also helped to minimize marketing risks. If one Edison product did not succeed in the market, there were other products to support research and manufacturing operations. The Edison Manufacturing Co.—a firm that Edison organized in December 1889 to produce phonograph batteries and telephone and telegraph systems—pursued this strategy in the 1890s, when it began making a number of Edison products, including fan motors, motion picture equipment, and medical devices.

In August 1887, Edison outlined this innovation strategy in letters to two potential investors: William Lloyd Garrison Jr., a Boston wool merchant, and James Hood Wright, a partner in the banking firm of Drexel, Morgan & Co. The West Orange laboratory, Edison explained to Garrison, "will be equipped with every modern appliance for cheap and rapid experimenting, and I expect to turn out a vast number of useful inventions and appliances in industry." Further, he

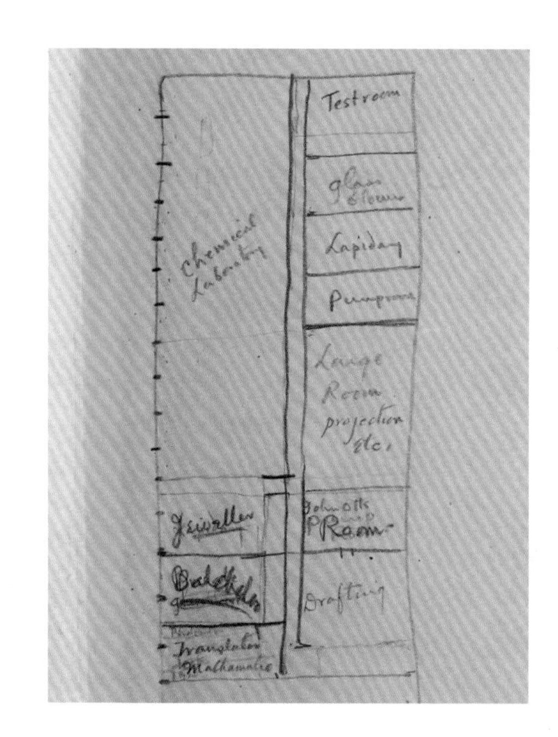

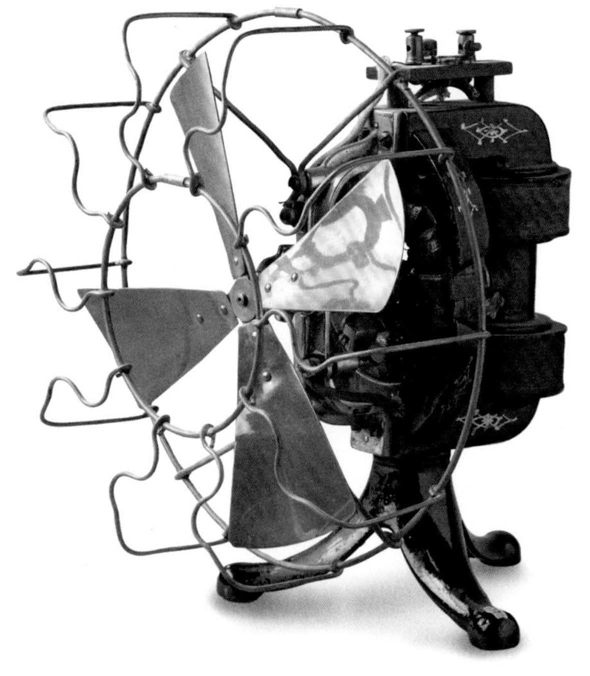

TOP: As workers constructed the West Orange lab, Edison drew a rough floor plan in an 1887 notebook. He reserved space for a glass blower, a jeweler, and two of his principal experimenters, Charles Batchelor and John Ott. He later moved the chemistry lab to its own building. BOTTOM: Electric fan produced by the Edison Manufacturing Co. in the 1890s.

told Wright that his "ambition is to build up a great industrial works in the Orange Valley starting in a small way & gradually working up." Using skilled workers, tools, machinery, and technical knowledge, Edison would produce inventions quickly and cheaply, mass-produce them in factories, and return any profits back to the laboratory to support new research.

Edison outlined some of his new ideas for Henry Villard, another potential financial supporter, in January 1888. Along with continued work on the electric light and phonograph, he planned to develop ore milling technology and to invent a variety of other products, including a hearing aid, a mechanical cotton picker, a coal sorting machine, a machine for clearing snow off of streets, and artificial silk, ivory, and mother of pearl.

The inventor wanted Garrison and Wright to invest in his company, "which shall have sufficient capital at its disposal to erect and equip factories for the manufacture of all my inventions" and market them through a network of wholesalers and retailers. Creating this company would give Edison the capital he needed to finance laboratory research and construct factories. "I honestly believe," Edison wrote to Wright, "I can build up a works in 15 or 20 years that employ 10 to 15,000 persons & yield 500 percent to the pioneer stockholders." Edison reminded Wright of his reliability: "You are aware from your long acquaintance with me that I do not fly any financial kites, or speculate and that the works which I control are well managed."

Wright politely put Edison off, telling him that he was going away on a trip and that he would talk to J. P. Morgan about the proposal when he returned. There is no evidence that Drexel, Morgan & Co. pursued the idea. Garrison expressed interest but doubted that Edison would find investors. "At present writing," Garrison explained, "owing to the peculiar state of the money market, it is not easy to interest capital in new enterprises . . . the general attitude of capital is that of caution."

By mid-September 1887, workers had completed half of the main building's roof and the foundations for two of the one-story satellite buildings. Charles Batchelor, Edison's close Menlo Park associate, came to West Orange in the fall to supervise the outfitting of the laboratory. While the workers continued construction, Edison and Batchelor scoured equipment and supply catalogs and filled notebooks with lists of chemicals, machines, tools, and supplies that Edison needed for the lab.

On November 25, Batchelor fired the boilers for the first time. On December 1 workers

The West Orange laboratory in the late 1880s.

began laying the conductors that would supply electricity to Glenmont. In his diary on December 23, Batchelor noted that they had "lit up Edison's house tonight for the first time from the laboratory." No one knows when Edison occupied the lab for the first time, but the first payroll was for the week ending January 5, 1888.

EDISON SPENT $180,000 ($4,309,000 today) constructing what he called "the best equipped & largest laboratory extant." He numbered the laboratory buildings one through five. Building 5—the main laboratory—housed a large research library, a stock room, two machine shops, a photography studio and darkroom, a lecture hall, and several experimental rooms.

Representing the forces driving the late-nineteenth-century industrial revolution—steam power, electricity, and information—Edison put the powerhouse at one end of the building and the library at the other. The steam boilers and generators in the powerhouse provided heat and electricity and powered the belt-driven machine tools in the laboratory's machine shops. An eighty-foot brick chimney stack rose above the powerhouse.

At the other end of Building 5 was the library—a large, ornate, high-ceilinged room that was paneled with varnished yellow pine. Above the main floor, where Edison and his secretaries kept their desks, were two balconies, divided into alcoves and filled with bookcases. Framed photographs, paintings, awards, and other mementos from Edison's career decorated the library's walls.

Edison used the library as a formal reception space. Mark Twain, who came to West Orange in 1888 to see Edison's improved phonograph, was an early visitor. Twain was fascinated by new technologies (he was one of the first major authors to use a typewriter) and eager to use the phonograph as a dictating machine. Edison later recalled that Twain spent several hours in the library recording funny stories on wax phonograph cylinders. These recordings, unfortunately, were lost in a 1914 fire.

Edison in the library of the West Orange laboratory, 1904.

In October 1909, another notable visitor to West Orange was Baron Eiichi Shibusawa, the president of the Dai-Ichi Bank of Tokyo and chairman of a commission representing the Chamber of Commerce of Japan. On April 12, 1912, Edison entertained Rudolf Diesel, a German engineer who invented the diesel engine in 1893. Diesel was traveling through the United States with a delegation of German scientists and engineers who were studying American industrial museums.

The library was also a source of information for Edison and his staff. Today the library contains about 10,000 volumes, including published patent reports, bound scientific and technical journals, and books on a wide range of subjects. To succeed as an innovator, Edison needed information—not only the latest developments in science and technology, but also data on the economy, sources of capital, equipment and supplies, and the activities of his competitors, business associates, and partners. Innovators with better sources of information enjoyed advantages over less-informed rivals.

Edison kept himself informed in several ways: He was a voracious reader of newspapers and magazines. He relied on credit reporting agencies to monitor the financial standing of business partners. His West Orange staff continued a practice, started at Menlo Park, of keeping articles relating to Edison's activities in scrapbooks. Later, in the early twentieth century, newspaper-clipping services routinely sent Edison-related articles to the laboratory, allowing him to track his image in the print media. Letters from business associates and consumers were another source of information, but keeping a well-stocked library was his primary means of keeping up-to-date on the newest scientific, technical, and business developments.

Across the hall from the library was the stock room, where Edison gathered the tools and materials his staff needed to develop new inventions. In 1927 Edison told Maurice Holland that the secret to inventing was to have a "big junk pile." He explained to James Hood Wright in 1887, "I shall carry a stock of almost every conceivable material of every size." Edison believed that maintaining a well-equipped stock room helped increase productivity by reducing the time experimenters would have to wait for supplies and equipment. "Inventions that formerly took months & cost a large sum can now be done in 2 or 3 days with very small expense," he wrote.

ABOVE: Stock room of the West Orange laboratory. Stock room clerks passed tools and supplies to machinists working in the heavy machine shop through this window. NEXT PAGE: Heavy machine shop, West Orange lab.

780

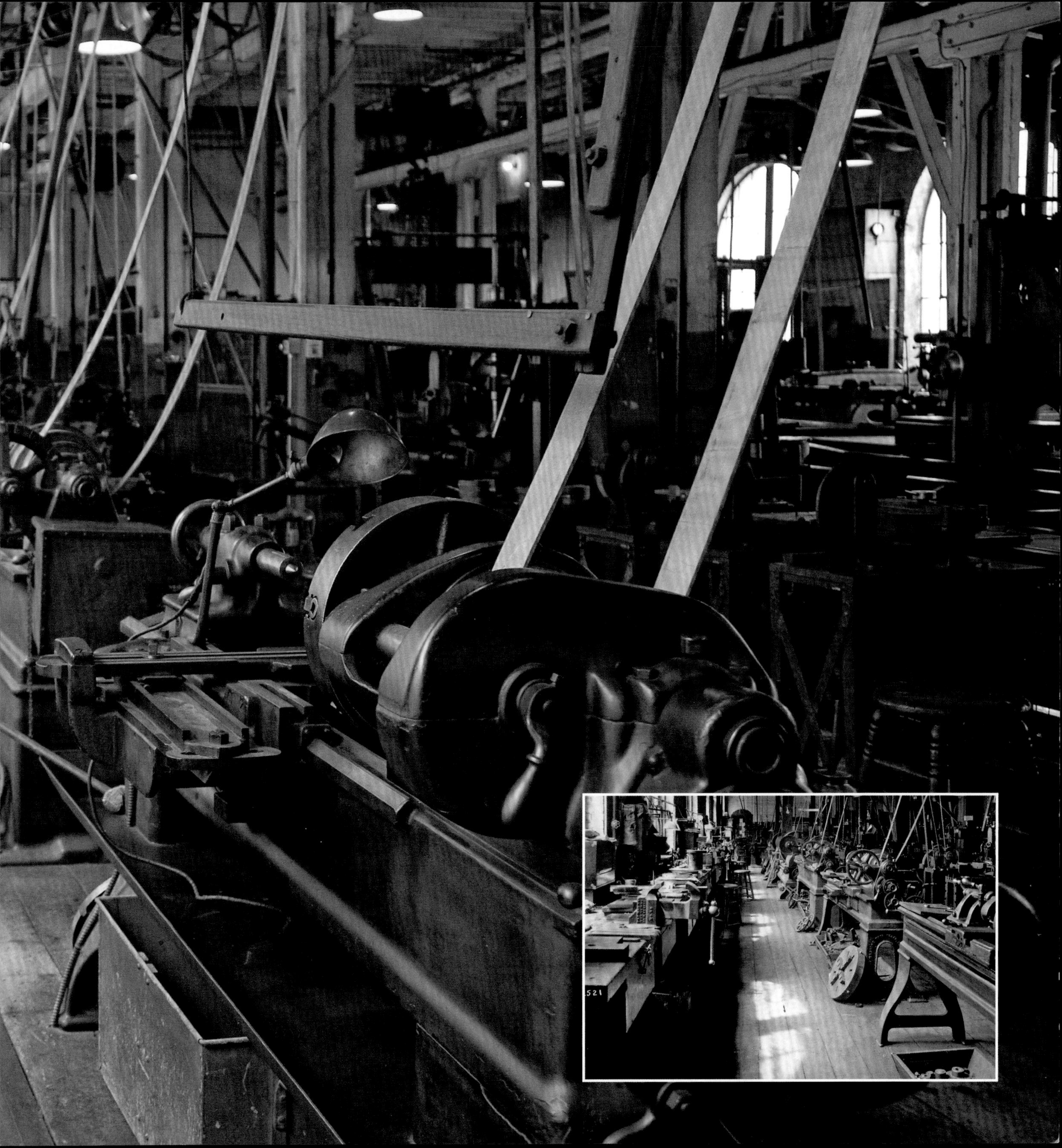

An order list Edison drafted in the fall of 1887 included two pounds of horse hair, two pounds of hog bristles, one ounce of seal hair, four ounces of porcupine quills, a mouse skin with the hair attached, ten pounds of walrus tusks, and two ounces of shark skin. He also ordered less exotic material: iron, copper, nickel, hard and soft rubber, and chemicals. Edison did not know if his experimenters would need hog bristles, but if they did, it was better to have them on hand. The initial expense of stocking the laboratory was high, but Edison concluded that he would save money in the long run. "With the latest machinery a man will produce 10 times as much as in a laboratory which has but little material."

Edison boasted to Wright that his laboratory was equipped to build everything from "a lady's watch to a locomotive." The heavy machine shop next to the stock room contained the machine tools capable of making the metal parts for large equipment, like steam locomotives. Large enough to employ fifty men, it contained drill presses, lathes, shapers, and milling and boring machines—all designed to cut, shape, and drill holes in metal. Rubber belts connected to shafts running along the ceiling on both sides of the room drove the machines. Two large direct current motors, mounted on a platform at one end of the shop, powered the two main shafts. An overhead traveling crane and six-ton chain hoist in the center of the room allowed machinists to easily move large metal pieces.

Precision machine shop on the second floor of Building 5.

Above the heavy machine shop on the second floor was the precision machine shop, which contained lathes, grinders, and milling machines designed for smaller, more delicate work, like ladies' watches. Today the precision machine shop extends from one side of the building to the other. When Edison opened the laboratory in 1887, a partition wall divided the room in half. The wall separated the precision machine shop from several experimental rooms, where Edison's staff worked on specific projects. In one of these rooms (no longer in existence), researchers conducted the first experiments on motion pictures.

At Menlo Park, Edison and his team worked together in one large room. The West Orange laboratory floor plan allowed Edison to assign workers to specific rooms, which gave experimenters more privacy. The proximity of experimental rooms to the machine shops and stock room, however, gave researchers easy access to tools and supplies, as well as the machinists who could translate their ideas into working models and prototypes. By taking down or moving partition walls, Edison could alter the lab's floor plan as his needs changed.

A door separated the precision machine shop from the rest of the second floor, which consisted of experimental rooms divided by a center hall. One room contained glass-blowing equipment for making experimental lamps, while another contained a mercury pump for creating vacuums in glass bulbs. Edison reserved Room 12 as his personal experimental space. Across the hall from Room 12, draftsmen took the rough sketches produced by Edison's experimenters and turned them into the precise, measured drawings that machinists used to construct invention prototypes.

The third floor of Building 5 included additional experimental rooms, exhibition

TOP: Room 12, Edison's private experimental room and the Drafting Room. BOTTOM: on the second floor of Building 5. The proximity of the machine shops and the drafting room to experimental rooms allowed Edison and his staff to easily communicate with one another.

space, a lecture hall, and a photographer's studio and darkroom. The lecture hall—included in Edison's original plan to demonstrate inventions to visitors and newspaper reporters—quickly became a music room, where researchers developed sound-recording techniques. In the early twentieth century, Edison used this room to evaluate sound recordings and audition musicians and singers for his record catalog. In the photography studio and darkroom, Edison's staff produced the images used to advertise Edison's products.

The one-story satellite buildings, positioned perpendicular to the main laboratory, were numbered one through four. Edison assigned specific functions or tasks to these buildings. Building 1, facing Valley Street, was the galvanometer room, which contained equipment for electrical research and testing. (A galvanometer is an instrument that measures electric current.) Because iron fittings interfered with magnetic instruments, this building was constructed with copper nails and brass pipes.

The galvanometer room reflected Edison's continued interest in electric light research at West Orange. Most of this research during the late 1880s focused on improving the design and performance of electric lamp filaments. Contract research Edison performed during the late 1880s and early 1890s for his electric light companies was an important source of income. In 1889 these companies—the Edison Electric Light Co., the Edison Lamp Co., the Edison Machine Works, and Bergmann & Co.—merged into the Edison General Electric Co. In 1890 Edison General Electric agreed to fund half of Edison's electrical experiments for five years. From 1890 to 1895, Edison received $121,000 ($3.3 million today) for research on direct current power systems and an alternating current motor. This contract remained in effect after April 1892, when Edison General Electric merged with a rival electrical manufacturer, Thomson-Houston Electric Co., to become General Electric. General Electric's support of the West Orange laboratory ended in the late 1890s,

Edison with Francis Arthur Jones in the chemistry lab, 1906. Jones, an English journalist, published a biography of Edison in 1908.

when it opened its own research laboratory in Schenectady, New York.

A courtyard separated the galvanometer room from Building 2, the chemical laboratory, which contained a long room with rows of experimental tables and storage cabinets filled with chemicals, test tubes, Bunsen burners, and other research equipment. At the back of the chemical lab was a smaller balance room, where chemists weighed and analyzed chemical samples. (Weighing samples in a separate room, a common practice in chemistry labs, helped prevent fumes and dust from contaminating the chemicals.) A slightly inclined cement floor allowed easy spill cleanup, and eight brick chimneys ventilated chemical fumes.

ABOVE: Edison in the chemistry lab, February 1908. In 1917, he told the *New York Sun*, "I have always been more interested in chemistry than in physics." NEXT PAGE: The West Orange chemistry lab, as it appears today.

Edison used the front section of Building 3 to store chemicals. The back section was a pattern shop, where carpenters made the wood patterns used to cast metal parts. The electric motors and belts that powered the shop's saws and other woodworking tools were mounted under the floor to allow workers to move around large pieces of wood. Edison sent the patterns to local foundries for casting, but because patternmaking required close collaboration among experimenters, machinists, and carpenters, he kept this function in the laboratory.

Building 4 contained a forge, blacksmith's shop, and metallurgical laboratory for analyzing and testing metals and ores, signaling Edison's plan to pursue ore milling research. During the 1890s experimenters in this building designed machinery for processing iron ore. In the early twentieth century, the function of Building 4 shifted to research on phonograph record–manufacturing techniques.

SILVER

ACID
SULPHURIC

AC/DC CONTROVERSY

In the summer of 1888, researchers at West Orange conducted electrocution experiments on animals. These gruesome tests were part of Edison's efforts to demonstrate the safety of his direct current (DC) electric power system and the dangers of George Westinghouse's alternating current (AC) system. Edison's system served an area of only one square mile. Alternating current, which distributed power at higher voltages and allowed power companies to serve larger areas, threatened Edison's electric lighting system.

An opportunity to challenge the Westinghouse system came in June 1888, when New York State enacted a law changing its method of capital punishment from hanging to electrocution. The law did not specify what type of electricity would be used, and the state asked the New York Medico-Legal Society to work out the details.

Edison encouraged New York to use Westinghouse equipment. He personally opposed capital punishment, but he believed that alternating current was the quickest way to end a human life. To meet the goals of the electrocution law—which required faster, more

In June 1888, George Westinghouse, Edison's competitor in the electrical industry, invited Edison to meet him in Pittsburgh to discuss cooperation between their companies. Edison politely declined.

efficient executions—experts needed more technical information, such as the amount of current and time required to kill a man, and the most effective way of wiring the prisoner to the equipment. To help New York State answer these questions, Edison offered the use of his staff and facilities at West Orange to perform electrocution experiments on animals.

From one perspective, the experiments were part of a cynical attack on a competitor. The physiological effects of electricity on living beings was poorly understood in the 1880s. Edison had a sincere concern about the safety of his electric power system, but he believed that his low-voltage DC system was safer than the AC system.

Over the course of the summer, Edison's electrical expert Arthur Kennelly electrocuted fifty dogs, ten calves, and two horses, carefully noting in a laboratory notebook the results of each experiment. Kennelly conducted the first experiment on July 12, when he electrocuted a thirteen-pound fox terrier. Wet bandages wrapped with bare copper wire were applied to the dog's legs. Kennelly applied a series of voltages, starting with 400 volts of DC and ending with a lethal dose of 1,000 volts.

No one knows where Edison got the animals. He had asked the American Society for the Prevention of Cruelty to Animals to send him dogs they were going to euthanize. The president of the ASPCA objected to the experiments.

The results of the electrocution experiments were tabulated and summarized in a report for the New York Medico-Legal Society. Not surprisingly, the report concluded that a 3,000-volt AC generator should be used in executions. A current of 1,000 to 1,500 volts applied for fifteen to thirty seconds was enough to cause death. In December 1888, the Medico-Legal Society accepted the report's recommendations, and New York State prison authorities adopted alternating current.

A new word was needed to describe execution by electricity. Edison suggested "ampermort," "dynamort," and "electromort." Edison attorney Eugene Lewis preferred "electricide," but he also suggested using Westinghouse's name. "As Westinghouse's dynamo is going to be used for executing criminals," he wrote, "why not give him the benefit of this fact in the minds of the public and speak of a criminal as being 'Westinghoused.'"

On August 6, 1890, William Kemmler, convicted of murdering his lover Tillie Ziegler with an axe, became the first man in the United States to die in the electric chair at Auburn State Prison in New York. Kemmler's execution was not quick and efficient, as the new capital punishment law intended, however. The first jolt of electricity failed to kill Kemmler, and a second seventy-second dose had to be applied.

Kemmler's botched electrocution defeated the purpose of New York's capital punishment law—to provide a more humane method of execution—but it was also a personal embarrassment to Edison. Ultimately, he lost the battle of the currents because AC was a more economical means of producing and distributing electricity. Cologne and Frankfurt were among the first cities in Europe to adopt AC electric power service in the early 1890s. In the United States, J. P. Morgan helped finance the Cataract Construction Co., organized in 1889 to generate hydroelectricity at Niagara Falls, New York. The company, which adopted the Westinghouse AC system, began servicing Buffalo in 1895.

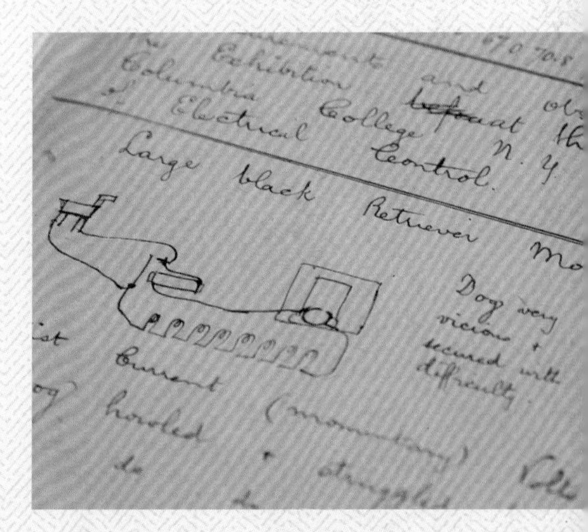

Experimenter Arthur Kennelly carefully recorded the results of the animal electrocution experiments in a notebook. After leaving the West Orange lab, he taught electrical engineering at MIT and Harvard.

DURING CONSTRUCTION OF THE LAB in 1887, Edison drafted an employee wish list, which revealed that he wanted a staff that could handle a wide range of technical problems. Included on this list were a watch toolmaker, a telegraph instrument maker, a glassblower, a mechanic who could work with tin and copper, draftsmen, and a mechanic who understood lapidary—or precious stone—work. Edison also wanted someone familiar with optics, a photographer, and "one person familiar with scientific matters to translate French, German and Italian into English."

The laboratory's first weekly payroll in January 1888 listed sixty-six employees who were paid collectively $376.05 (about $9,170 today). Most of these workers were laborers, carpenters, and machinists who were busy "fitting up" the laboratory, but the payroll also listed experimenters who worked on the phonograph, ore milling, electric lamp filaments, telegraph systems, and wire insulation.

The number of employees fluctuated as Edison's requirements changed. When the laboratory was busy, he added more experimenters to the payroll. During slack times he reduced

Edison (front row, center) and the West Orange laboratory staff in 1893. Charles Batchelor is to Edison's left, and John Ott is to Edison's right. William K. L. Dickson, the principal experimenter on motion pictures, stands behind and slightly to the left of Batchelor.

the staff. The laboratory payroll for the week of June 12, 1890, for example, included fifty employees, who together were paid $805.22 (about $20,500 today). During the week of March 26, 1891, Edison employed sixty-nine workers, paid collectively $1,072.52 (about $27,400 today) to work on a variety of experiments, including ore milling, electroplating, wire insulation, phonograph record duplication, and street railway motors.

For the week of August 31, 1893, there were only forty-four workers on the laboratory payroll, including nine experimenters, seven machinists, eight draftsmen, two laborers, a patternmaker, a carpenter, and a part-time typist named Miss Mandeville. Edison's experimenters—the highest-paid lab employees—were paid by the week. John F. Ott, the lab's superintendent, earned $34.60 that week ($893 today). Each of the following workers earned thirty dollars ($774 today): John Ott's brother Fred, who was one of Edison's principal experimenters; William K. L. Dickson, who worked on ore milling and motion pictures; and Arthur Kennelly, who worked in the galvanometer room. Other occupations were paid by the hour. Edison paid machinist Charles Hopflinger 32.5 cents per hour, and for the sixty hours he worked during the week of August 31, he received $19.50 ($503 today). Draftsman Thomas Banks earned 67.5 cents per hour, while Miss Mandeville, who worked only forty hours that week, received $6.67.

BY THE EARLY TWENTIETH CENTURY, Edison had created an industrial works at West Orange. The laboratory stood at the center of a manufacturing complex that produced, among other things, motion picture equipment, phonographs, sound recordings, Portland cement, and storage batteries. Before 1910, these products had been manufactured and sold by separate companies. The Edison Manufacturing Co., for instance, made motion picture projectors and films, electric fans, and X-ray machines. The Edison Storage Battery Co. controlled storage batteries; the Edison Portland Cement Co. administered Edison's cement business; the Edison Phonograph Works manufactured talking machines and records; and the National Phonograph Co. marketed them to consumers.

The laboratory continued to experiment on new technologies, but it also worked to improve existing Edison products and designed the tools and machines used to mass-produce factory products. Edison's manufacturing and marketing companies, in turn, generated the profits that he needed to keep the laboratory in operation.

The informality of the Menlo Park and early West Orange periods—when Edison's closest laboratory associates could easily move from the lab to company management positions—was replaced in the early twentieth century with a corporate bureaucracy defined by organizational charts. As always, Edison remained in control of his companies and had the final word on all major policies, but the scale and scope of the West Orange operation in the early twentieth century required him to delegate more responsibility to his company managers. It was still possible for talented lab associates to work their way up into managerial positions without professional experience, but a number of younger executives in the Edison organization had university training. For example, William Maxwell, who joined Edison in 1911 as a phonograph sales manager, had attended the University of Virginia. In 1915 he became general manager of the Edison Co. Phonograph Division. Stephen Mambert, a Cornell University–educated management engineer, began working for Edison as a cost accounting clerk in 1913. By 1916 he was a vice president responsible for implementing a plan to reduce financial costs.

Carpenters working in the pattern shop in January 1917.

Frank L. Dyer, a patent attorney who became Edison's general counsel in 1903, also brought managerial and administrative skills to the Edison organization. Before becoming the president and general manager of the National Phonograph Co. in 1908, Dyer handled Edison's complicated legal and patent affairs, including several contentious lawsuits that threatened the phonograph and motion picture businesses. In 1910 Dyer convinced Edison to approve a plan to consolidate his various companies into one corporate entity, Thomas A. Edison, Inc. Up to that point, Edison had used the profits from his successful phonograph and film businesses to support struggling ventures like the storage battery and Portland cement. Moving funds from healthy companies to struggling ventures allowed Edison to keep unsuccessful projects alive, but it also deprived the more profitable companies of capital for expansion. Dyer proposed merging the companies to reduce administrative expenses by centralizing accounting, legal, advertising, and other managerial functions. Another advantage, noted by Dyer—who coauthored Edison's official biography, *Edison: His Life and Inventions,* in 1910—was to "bring home more firmly to the public that this is Mr. Edison's personal business and that his personality stands behind it."

According to folklore, lab workers placed this sign on the elevator in Building 5 as a joke, knowing that Edison preferred using the stairs.

WHEN EDISON OPENED the West Orange laboratory at the end of 1887, he was already a world-renowned inventor who had made significant contributions in several industries. If he had done nothing else for the rest of his life, history would have remembered Edison for his work in telegraphy and the telephone, the invention of the phonograph, and the development of his incandescent electric lighting system. At age forty, however, Edison was still relatively young, and he was not content to rest on his laurels. Equipped with a steady stream of ideas, Edison worked at West Orange for the remaining forty-four years of his life—improving his phonograph; inventing a motion picture camera; developing technologies to process iron ore and manufacture Portland cement; building an improved storage battery; conceiving a method for constructing cement houses; and even attempting to cultivate a domestic source of natural rubber. Edison also became a manufacturer and marketer of his products and spent much of the First World War conducting research for the U.S. Navy. The remaining chapters discuss Edison's "second act" as an innovator.

Edison
A PRODUCT OF
THE EDISON
LABORATORIES
51741-L
IT DON'T DO NOTHING BUT
RAIN
(Phil Cook)
BILLY JONES AND
ERNEST HARE
(THE HAPPINESS
BOYS)
10925

A PHONOGRAPH IN EVERY HOME

"IT IS AN EASY MATTER TO GET SOME MEN TO . . . PRODUCE GOODS, BUT IT REQUIRES A CONSIDERABLY HIGHER TYPE OF MAN TO SUCCESSFULLY SELL THE GOODS."

AT MENLO PARK Edison unsuccessfully attempted to turn the tinfoil phonograph from a scientific curiosity into a commercial product; at West Orange he succeeded. In the late nineteenth century, Americans began producing and consuming a steady stream of mass-marketed goods. Edison's efforts to design, manufacture, and market phonographs at West Orange from 1887 to 1929 were a part of this market transformation.

Pages 96–97: Edison's Diamond Disc phonograph in the West Orange lab music room. ABOVE: A pajama-clad Edison listening to a disc phonograph record at Glenmont, September 1912.

Mass consumer marketing in the late nineteenth century was made possible by several changes in the way the economy manufactured, sold, and consumed products. New manufacturing techniques allowed companies to increase production, while railroads and the telegraph enabled local businesses to expand their operations into national markets. To promote their products, manufacturers adopted new marketing strategies, including branding and mass-circulation advertising. Producing goods in high volume and selling them at low unit prices—a key mass-marketing strategy—made products more affordable to consumers.

At the same time, rising living standards and changing attitudes about spending and leisure encouraged consumption. Cultural values emphasizing leisure and immediate gratification replaced an older ethic that stressed hard work and self-sacrifice. As a result, late-nineteenth-century Americans embraced a variety of new leisure activities and began purchasing manufactured products that were previously made in the home, like soap and clothing. The American household shifted from a center of production to one of consumption.

At first, Edison viewed the phonograph as an office dictating machine, believing that there would be high demand for dictating machines, especially from businesses in areas lacking access to trained stenographers. The steady increase in the value of office equipment production, from $3.6 million in 1879 to $8.2 million in 1889 (from $83.7 to $207 million today), suggested that he was correct. In October 1878, Edison spoke about office workers dictating up to 4,000 words on a tinfoil phonograph. When he resumed phonograph research in October 1886, shortly before beginning construction of the West Orange laboratory, an experimenter noted Edison's plan to design "a small compact instrument, suitable for office use."

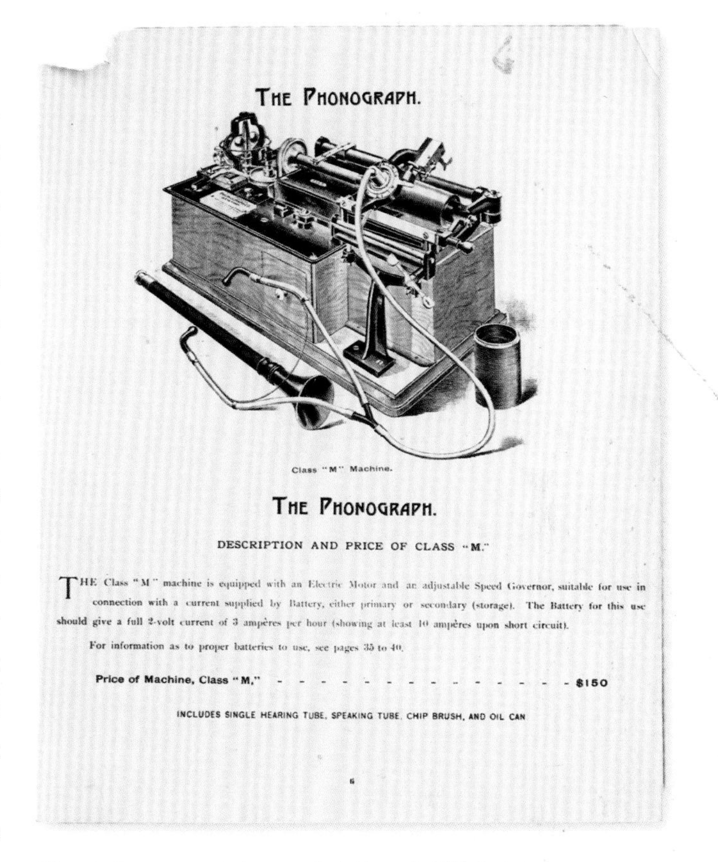

THE PHONOGRAPH.

Class "M" Machine.

THE PHONOGRAPH.

DESCRIPTION AND PRICE OF CLASS "M."

THE Class "M" machine is equipped with an Electric Motor and an adjustable Speed Governor, suitable for use in connection with a current supplied by Battery, either primary or secondary (storage). The Battery for this use should give a full 2-volt current of 3 ampères per hour (showing at least 10 ampères upon short circuit).

For information as to proper batteries to use, see pages 35 to 40.

Price of Machine, Class "M," - - - - - - - - - - - - - - **$150**

INCLUDES SINGLE HEARING TUBE, SPEAKING TUBE, CHIP BRUSH, AND OIL CAN

6

Edison battery-powered phonographs in the early 1890s included features that allowed office workers to record dictation.

In the fall of 1887 Edison completed the design of his "improved" phonograph. Unlike the hand-cranked tinfoil phonograph, this model was a battery-operated machine with a removable wax cylinder, a recorder/reproducer assembly, a cutting tool for shaving wax cylinders, and a stop/start control. The improved phonograph was significantly better than the tinfoil machine, but it was more of a prototype than a commercial product.

Edison and his staff continued phonograph research into 1888. Edison worked on mechanical features, while Arthur Kennelly focused on the electric motor and battery. Jonas Aylsworth, a chemist hired in October 1887, developed wax cylinder compounds. He conducted more than 700 experiments on waxes, soaps, and fatty acids to produce a cylinder soft enough to take a recording yet durable enough to withstand repeated reproduction. A. Theo.

Wangemann experimented on recording techniques, and Franz Shulze-Berge developed record duplication methods. On April 30, 1888, Edison organized the Edison Phonograph Works, and in June 1888 he unveiled a modified version of the new machine, known as the "perfected" phonograph.

While Edison had outsourced phonograph production at Menlo Park, he maintained control of phonograph manufacturing at West Orange. He planned to make phonographs using the "American System of Manufactures," a production process introduced in the early nineteenth century to manufacture firearms. Instead of skilled craftsmen making individual parts, unskilled laborers assembled phonographs with prefabricated parts. This system lowered production costs, increased productivity, and resulted in uniform, standard products. Lower production expenses enhanced Edison's competitiveness by allowing him to decrease costs to consumers. Keeping phonograph production close to the laboratory also allowed Edison and his experimenters to rapidly implement design changes and quickly introduce improvements in manufacturing techniques. Thus, phonograph production increased from ten machines a day in the fall of 1888 to fifty by the spring of 1889.

Window display for Edison wax cylinder records.

Edison sold the phonograph marketing rights to Jesse Lippincott, a Pittsburgh glass manufacturer who had earlier purchased the rights to the graphophone, a rival talking machine invented at the Washington, D.C., laboratory of Alexander Graham Bell. Lippincott organized the North American Phonograph Co., which licensed subsidiaries in exclusive territories to lease phonographs. By May 1890 there were thirty-two licensed local phonograph companies in the United States. The North American Co. leased phonographs to the local companies for an annual fee of $20 ($510 today). The local companies, in turn, leased phonographs to customers for $40 per year.

Typists transcribing letters from Edison phonographs. The expansion of office workers in the 1880s and early 1890s convinced Edison that there would be a large market for dictating machines.

Technical and marketing problems prevented the North American Phonograph Co. from establishing a profitable talking machine business. Changes in temperature and humidity warped and cracked the wax cylinders, making them difficult to remove from the machine. Cylinder records could be shaved for reuse, but the shavings clogged the machine.

Customers also found the phonograph too complicated. One local phonograph company manager complained that office workers did not have the technical skills to operate the machine. Another manager told Edison "that if the entire mechanism could be rearranged, made much lighter and more compact it would be desirable. The present phonograph is much too heavy and is liable to create an impression that it is exceedingly complicated."

The most serious technical problem concerned the phonograph's chemical battery. Battery-powered electric motors gave the phonograph cylinder steady and even rotation, but customers had trouble getting the batteries to work properly. One user told Edison's secretary, "I followed the directions closely and short-circuited the battery about fifteen minutes after it had been set up almost 12 hours. The battery seems to 'run down' after it has been running

TALKING DOLL

Edison's talking doll, developed at West Orange during the late 1880s, was part of his strategy of quickly designing new products in the laboratory, manufacturing and selling them on a mass scale, and returning the profits to the lab to support new research.

In October 1887, Edison signed a contract with William Jacques, an employee of the Bell Telephone Co. in Boston, and his associate Lowell Briggs to market talking dolls. Jacques and Briggs, the inventors of a doll phonograph mechanism, then organized the Edison Phonograph Toy Manufacturing Co. Edison was dissatisfied with their phonograph mechanism design, however, and asked Charles Batchelor to improve it.

In February 1888, the laboratory designed a sound reproducing mechanism small enough to fit inside a doll, and on March 6, 1888, Batchelor wrote in his notebook, "Made a small phonograph for dolls, etc. with an automatic return motion so that you simply turn always in one direction and it says the same thing over and over again."

A rough sketch shows that experimenters considered placing the phonograph inside the doll's head, with the speaker horn placed near

Following Edison's strategy of rapidly developing new products, the West Orange lab quickly introduced the talking doll, but it failed to meet consumer expectations.

the doll's mouth and a pull string at the back of the head, but in the final design they mounted the mechanism inside the doll's torso.

In the fall of 1888, newspapers began publishing stories about Edison's doll. As the *New York Sun* reported on November 22, "Children all over the world will before long have reason to bless the name of Thomas Edison for that wizard has just perfected a toy the like of which was never dreamed of by them even on Christmas Eve." Edison, the *Sun* announced, would begin manufacturing and shipping dolls in large quantities so that, soon, "children not only in America, but also in Europe, and even in far off Russia will be able to possess dollies that in their owner's native language can talk to them." Edison hoped to have the dolls ready for the 1888 Christmas trade, but, as his secretary told the editor of the *New York World*, "the details of the mechanism have not been satisfactorily arranged."

Because of design problems and production delays, Edison missed the 1889 Christmas trade and did not begin producing dolls until early 1890. By March 1890, the Edison Phonograph Works, which manufactured the mechanisms and inserted them in doll bodies purchased from Germany, had manufactured 3,000 dolls. Toy dealers began selling the dolls, but, because the mechanisms were delicate and easily damaged, consumers soon began returning them. As one dealer reported, "We are having quite a number of your dolls returned to us. . . . We have had five or six recently sent back some on account of the works being loose inside, and others won't talk and one party sent one back stating that after using it for an hour it kept growing fainter until finally it could not be understood." Edison took the talking doll off the market in April 1890.

Edison National Historical Park preserves the earliest surviving talking doll recording. On the recording, produced in 1888, an unknown woman recites the nursery rhyme "Twinkle, Twinkle, Little Star." Museum curators found the thin, ring-shaped metal recording in Edison's library in 1967, but, because it was significantly twisted, it could not be played on sound recording equipment. In 2011, scientists at the Lawrence Berkeley National Laboratory created a digital model of the recording and reproduced the audio with optical scanning technology.

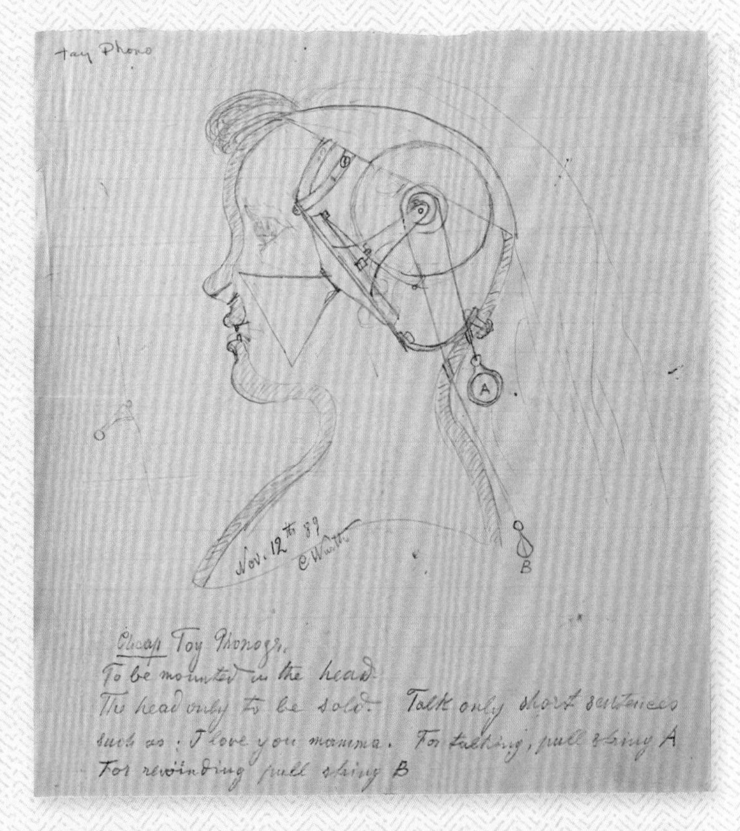

In this November 1889 drawing, experimenter Charles Wurth envisioned a phonograph mechanism in the talking doll's head. Ultimately, Edison put the mechanism in the doll's body.

the machines about five minutes and then if the machine is stopped the battery will not start it again without assistance." As Louis Glass, the general manager of the Pacific Phonograph Co., reported to Edison, "The batteries received with machines do not thus far seem to be altogether satisfactory. A single cell will run some instruments up to 160 revolutions whilst others are barely moved and we cannot get 100 revolutions." Customers complained about the battery's high cost and messy maintenance schedule. Phonograph batteries were primary cells, requiring the electrodes and chemical electrolytes to be changed when they were depleted. The Michigan Phonograph Co. tersely telegraphed Edison: "Phonograph batteries worthless."

AS THE LOCAL PHONOGRAPH COMPANIES struggled to market office dictating machines, the phonograph succeeded as a source of entertainment. In the fall of 1889 Louis Glass invented an attachment that changed the phonograph into a coin-operated machine. On November 23, 1889, Glass's Pacific Phonograph Co. installed a coin-operated phonograph in San Francisco's Palais Royal Saloon. This phonograph was equipped with a musical cylinder record and rubber listening tubes for up to four customers. For a nickel, saloon patrons could hear a recorded song, short comedy skit, or dramatic reading. On December 4, a second coin-operated phonograph was installed in the saloon, and a third was delivered on December 10. The company's fourth machine was installed in the waiting room of the Oakland–San Francisco ferry. By May 1890 the Pacific Co. operated fifteen coin-operated machines throughout San Francisco.

Between November 1890 and January 1891 the phonograph companies installed 704 coin-operated phonographs in saloons, hotels, and railroad stations across the country. These coin-operated amusement phonographs were highly profitable. San Francisco's first coin-operated machine earned $1,035.25 from November 1889 to May 1890 ($26,400 today). The Missouri Phonograph Co. made $1,500 a month from its forty-eight machines ($38,300 today). The Eastern Pennsylvania Phonograph Co. generated up to $550 per month from its twenty-five machines ($14,000 today). Virginia's Old Dominion Phonograph Co. reported, "In towns of three, four, five, six hundred and a thousand inhabitants the business pays. We can place the instrument in those little towns and they serve as a plaything, as a variety show. The people in these towns have only the church to go to—they cluster around the machines."

Despite its success, some phonograph business leaders were reluctant to promote amusement phonographs. North American Phonograph Co. general manager Thomas Lombard

LEFT: The popularity of coin-operated phonographs installed in train stations, hotels, and other public places in the early 1890s changed the focus of the industry from business communications to entertainment. RIGHT: Before Edison began selling horns with his spring-driven phonographs in 1896, listeners used rubber tubes and earpieces to hear phonograph records.

urged local companies not to rely on coin-operated machines. "The 'coin-in-the-slot' device is calculated to injure the phonograph . . . as it has the appearance of being nothing more than a toy, and no one would comprehend its utility as an aid to businessmen and others for dictation purposes." Virginia McRae, editor of the talking machine industry trade magazine *The Phonogram,* predicted that "the phonograph will soon cease to be the toy of the people who frequent saloons and cigar shops."

The profitable coin-operated amusement phonograph, combined with the commercial failure of the dictating phonograph, changed the focus of the sound recording industry. In June 1893 Edison admitted, "Our experience here shows that a very large number of machines go into private houses for amusement purposes—that such persons do not attempt to record nor desire it for that purpose."

Because Edison was the North American Co.'s largest creditor (the company had not paid him all of the money it owed for marketing rights or experimental expenses), he forced the company into receivership. Under this procedure, a court-appointed receiver inventoried the company's assets and liabilities, then auctioned the assets to pay off any creditors. When the receiver sold the company's assets in early 1896, Edison was the successful bidder and regained control of his phonograph patents.

On January 27, 1896, Edison organized the National Phonograph Co. to market less expensive, spring-driven cylinder machines. In December 1896 he introduced a lightweight spring-powered machine called the Home phonograph, priced initially at $40 and lowered to $30 in 1897 (approximately $1,110 and $839, respectively, today). The development of a reliable spring motor, which provided steady rotation to the phonograph's cylinder, in the late 1890s helped make the phonograph more attractive to consumers by eliminating the need for complicated batteries.

The Home phonograph was the first in a line of affordable talking machines the National Phonograph Co. marketed to consumers for entertainment purposes. In 1898 Edison unveiled a lighter and less expensive cylinder machine, the Standard, which weighed seventeen pounds and sold for $20 ($560 today). In 1899, he introduced an even lighter and cheaper machine, the Gem, which weighed seven and a half pounds and retailed for $7.50 ($210 today). Lower machine costs allowed National to increase phonograph sales. Machine sales climbed from 1,200 in 1896 to more than 113,000 in 1903. Record sales increased from zero in 1896 to seven and a half million in 1904.

The development of less expensive entertainment phonographs in the late 1890s represented an important change in the way consumers used sound recording technology. Edison had eliminated the complicated recording controls found on earlier dictating phonographs, which made them less expensive to produce and less complicated to operate. It also shifted the responsibility of making recordings from consumers to talking machine manufacturers. Consumers could purchase attachments that allowed them to make home recordings, but Edison's entertainment phonographs were designed mainly for play-back only.

BY THE EARLY 1900S Edison was a leading manufacturer of entertainment phonographs, but he faced competition from the disc talking machine manufactured by the Victor Talking

The Edison Fireside phonograph, introduced in July 1909 and encased in a mahogany cabinet, sold for $39 ($995 today).

Ads for Edison entertainment phonographs featured colors and images designed to appeal to consumers. This marked an important shift in advertising from the simple black-and-white text and graphics of earlier, business-related Edison products.

Machine Co. Organized in 1901 by Camden, New Jersey, machinist Eldridge R. Johnson, the Victor Co. produced the gramophone that was invented by Emile Berliner in the late 1880s. Johnson expanded gramophone sales by designing a tone arm that improved the machine's acoustic quality. In 1906 he unveiled the Victrola, a disc player with an enclosed speaker horn encased in an attractive wood cabinet.

Victor's disc records were easier to store than Edison's cylinder records and could be played longer—from four to seven minutes—allowing the company to record music that would not fit on Edison's two-minute cylinder records. Further, Victor's record catalog featured Adelina Patti, Nellie Melba, Enrico Caruso, and other celebrities that it promoted through extensive advertising. These strategies helped Victor increase machine sales from 7,600 gramophones in 1901 to 98,000 in 1907.

Italian tenor Enrico Caruso and the Victrola phonograph, ca. 1910. Recording celebrities like Caruso allowed the Victor Talking Machine Co. to become Edison's chief competitor in the entertainment phonograph business.

Edison's first response to Victor focused on technology. To compete with Victor's longer disc record, Edison introduced a four-minute cylinder record called the Amberol in 1908. He also developed a new cylinder machine—the Amberola—which mimicked the Victrola in both name and appearance. The Amberol record improved sales, but it did not challenge Victor. Consequently, Edison's business managers convinced him to develop a disc machine and record. In 1912, Edison introduced the Diamond Disc phonograph, which featured a diamond-pointed reproducer and a record made out of a material called condensate. That same year he began marketing the Blue Amberol, a celluloid cylinder record that produced a much higher sound quality than earlier cylinders.

THE EDIPHONE

In 1903, as the National Phonograph Co. expanded the market for entertainment phonographs, the West Orange laboratory resumed work on an office dictating machine. Unveiled in 1905, the Edison Business Phonograph allowed office workers to dictate letters onto a wax cylinder, which stenographers could then replay to type a paper letter for mailing.

In 1912 the company renamed the business phonograph the "Edison Dictating Machine." Between 1911 and 1914, the laboratory's engineering department designed an integrated office dictating system, which included an executive model, a secretarial machine, and a shaving machine to recycle used cylinders. Laboratory engineers also designed the Telescribe, which recorded telephone conversations on a phonograph cylinder. Introduced in 1914, the Telescribe did not perform well in noisy offices. In 1918 it was improved and reintroduced. That same year, Edison's advertising department renamed the "Edison Dictating Machine" the "Ediphone"—a less cumbersome name that evoked its inventor and more closely resembled the name of a competing machine, the Dictaphone, which was becoming a generic term for office dictating equipment.

During the 1920s, the West Orange laboratory continued to improve the Ediphone's design, making it simpler, smaller, and less expensive. As Edison claimed in a late 1920s advertising booklet, "I have spared no effort in developing the Ediphone because ambitious people use it and they deserve the best assistance."

The company continued manufacturing Ediphones after Edison's death in 1931. In 1951, the company redesigned the Ediphone again and introduced the Edison Voicewriter, producing a portable version the following year. Thomas A. Edison Inc. and its successor, McGraw-Edison, manufactured Voicewriters into the mid-1960s.

SYSTEM for FEBRUARY—ADVERTISING SECTION

THE AMERICAN DICTATOR

Edison Dictation has become a National habit. Business firms by thousands are solving their office problems with this short-cut from brain to type.

Simply say *YOU* are ready to begin and the Edison Service Organization, everywhere, will introduce you to the

EDISON DICTATING MACHINE

Please mention SYSTEM when writing to advertisers

Despite Edison's failure to develop an office dictating machine market in the 1890s, he reintroduced his business phonograph in 1905.

In the early twentieth century, the major phonograph companies—Edison, Victor, and Columbia—sold both the talking machines and the records to play on them, so they needed record catalogs that appealed to consumers. Predicting which recording artists and songs would sell, however, was difficult. Victor adopted a successful approach: attract consumers by recording celebrity artists.

Edison refused to use celebrities to promote his records. "We care nothing for the reputation of the artist, singer or instrumentalist," he wrote in 1912. "All that we desire is that the voice shall be as perfect as possible." In a November 1911 letter explaining his recording policies, Edison noted, "There are, of course, many people who will buy a distorted, ill-recorded scratchy record if the singer has a great reputation, but there are infinitely more who will buy for the beauty of the record, with fine voices, well instrumented and no scratch."

Edison based his marketing strategy on the belief that his records and phonographs were technologically superior. Consumers, as a result, would buy his products when they heard their finer sound quality. "It is not our intention to feature artists or sell the records by using the artist's name," Edison proclaimed. "We intend to rely entirely upon the tone and high quality of the voice."

In 1911, Edison began listening to the records in his own catalog, and he did not like what he heard. "We use bands when they should be orchestras. We keep instruments in our orchestra which hurt the whole by beating and interfering with the better instruments. We accompany a singer with the other instruments. We accompany a singer with a loud strident blast, when it should be soft and mellow. Our men play out of time, they do not time well."

To compete with the Victrola, in 1909 Edison introduced the Amberola, a cylinder phonograph with an enclosed horn priced at $200 ($5,100 today).

Edison also listened to Victor recordings. Overall, he was not impressed, noting in May 1912, "All of the Victor records up to this point are very weak on the table machine three feet away. I can scarcely hear anything but strong notes. The scratch is bad in high priced records & very little in low priced." He listened to Victor's recording of "Crucifix," a duet featuring Enrico Caruso and the famous French opera singer Marcel Journet. Edison liked the tune and thought Caruso's accompaniment was fair, but he did not think Caruso and Journet were a good combination. He also detected a *tremolo*, or trembling effect, in their voices that he disliked intensely.

Two additional Victor recordings Edison listened to featured Al Jolson, who later starred in the first feature-length sound motion picture, *The Jazz Singer* (1927), and became one of the highest-paid entertainers of the 1930s. In regard to Jolson's "That Haunting Melody," recorded on December 22, 1911, Edison noted "abnormally sharp voice, metallic—tune not good." He liked Jolson's"Snap Your Fingers" even less, simply jotting in his notebook, "Coney Island beer saloon singer—not for us."

Dissatisfied with his record catalog and armed with definite opinions about the music he liked and disliked, Edison effectively became his company's music director. He spent hours in the laboratory music room evaluating and selecting the artists and songs the company would record. He filled notebook after notebook with comments and directions for the music room staff.

Edison liked baritone Francis Rogers, who sang "When the Roses Bloom" and "Little Irish Girl" on January 14, 1915. "Very good, mellow, even voice. You can use him—not for the Irish girl type of song—those of the Roses Bloom type." He was less confident about Harry Williamson. "This man has bad tremolo most notes—Is not as good as tenors we have. However as public want new singers we might take a few from him if not too expensive."

The Edison Co. produced ten recordings by Russian composer and pianist Sergei Rachmaninoff, but Edison disliked his style (he called him a "pounder") and refused to release more records. In 1920 Rachmaninoff signed a recording contract with Victor.

Edison approached music selection as a technical problem, not a marketing issue. He deconstructed music like a machine, identifying the parts that he liked and those he disliked—an analogy he made himself, in 1911, when he wrote to his European representative Thomas Graf, "We have a flute that on high notes gives a piercing abnormal sound like machinery that wants oiling."

Edison expected consumers to appreciate his focus on technical quality, but not everyone cared. Edison's Omaha, Nebraska, dealer received one complaint: "Let's have more attention paid to the spirit of a song rather than to make certain that every solitary instrument is surely heard. . . . If the public cared anything for absolute perfect recording where every instrument is clearly heard, the Victor would have been out of business long ago."

As the foxtrot and other new dance steps became popular in the 1920s, young phonograph users began changing the machine's speed to increase record tempo. Edison was appalled. In 1925, the seventy-eight-year-old inventor asked one of his engineers to modify the phonograph so that consumers could not change the record speed. "This change of speed is far worse than any loss due to having dance records too slow. They are not too slow they are absolutely right time but young folks of the family want this fast tune. . . . I don't want it & won't have it. It does a great injury in more ways than one."

Edison listening to mezzo-soprano Helen Davis, accompanied by pianist Victor Young, in the West Orange lab music room, ca. 1926.

Dealers complained when the Edison factory removed the phonograph's speed regulator. "Mr. Edison is still working along the plan of trying to sell the public what he thinks is good for them, rather than what the public wants," the Kent Piano Co. wrote. "We fail to see why a customer who has just spent his good dollars for a phonograph should be denied the privilege of playing the latest dance records at the speed he has been accustomed."

"KEEP ON BUILDING CASTLES IN THE AIR"

PERCY WENRICH'S LATEST AND GREATEST SONG

Edison Re-Creation No. 51016

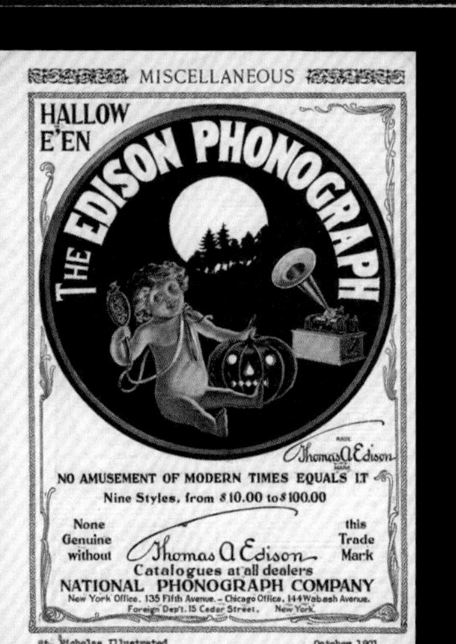

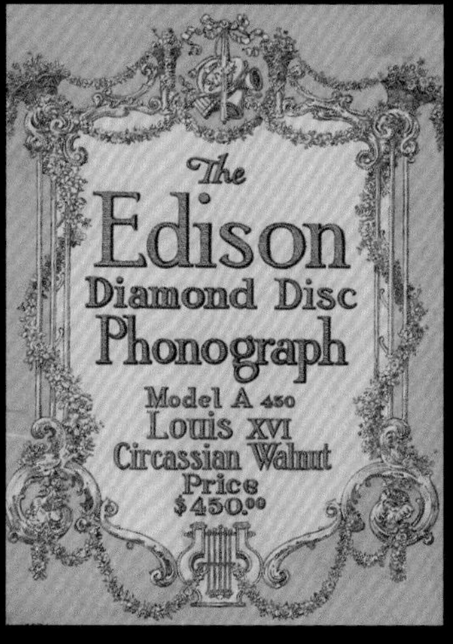

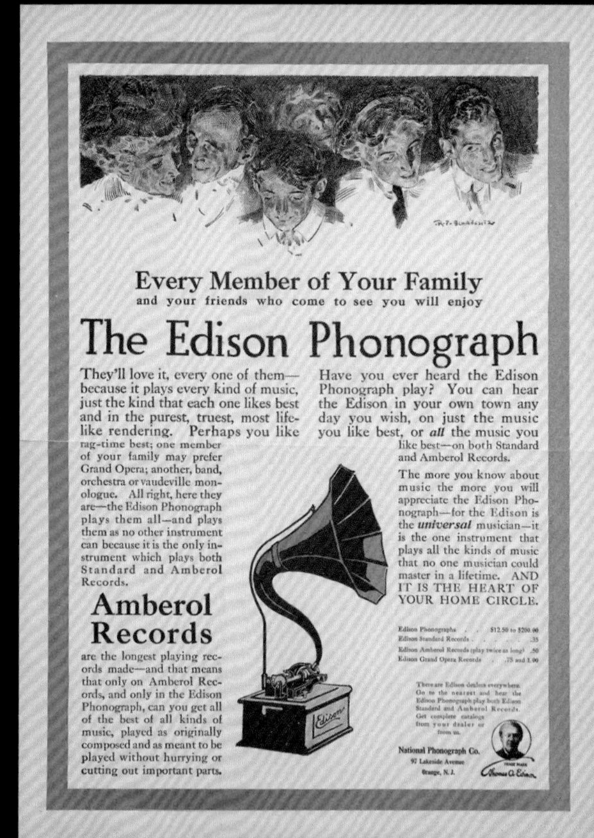

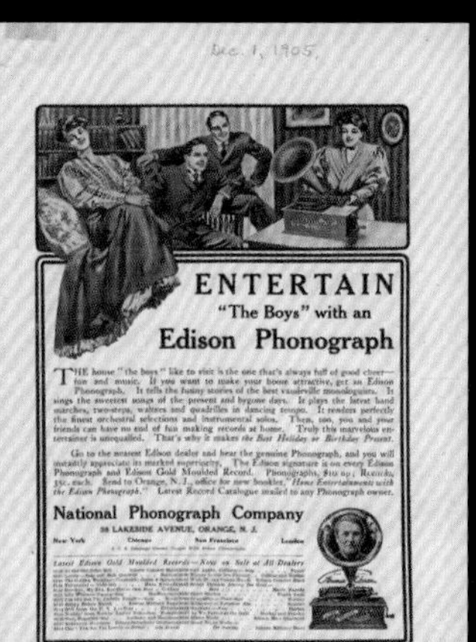

Ads for Edison phonographs and records. Unlike record labels of today, Edison and his sound-recording competitors in the early twentieth century produced both records and record players.

Edison's personal taste in music was one problem; another was the company's inability to respond quickly to rapidly changing consumer preferences, particularly during the First World War, when more modern forms of music like jazz became popular. Except for letters from individual customers and dealers, the company lacked comprehensive market data to determine what the public wanted. Finding a successful balance between classical and more popular music was also difficult. If the company produced too much classical music, some consumers complained about the lack of popular tunes. If it recorded too much jazz, it would get letters like the one written in 1926 by H. E. McConnell, who complained, "For God's sake give us more real music like that you have recorded instead of so damned much foolish and unbearable jazz."

Victor succeeded in part because of its advertising strategy. Edison was not opposed to mass-circulation advertising, even for the phonograph, but he did not spend as much money on ads as his competitors because he believed they could not convey the technical superiority of his products. The goal of his marketing strategy, therefore, was to allow as many consumers as possible to hear his phonographs and records. "Hearing the tone of the Edison," the company claimed in one ad, "creates the desire to own one and produces a natural comparison in the hearts of everyone hearing the instrument and any needle talking machine they ever hear."

Edison tested this marketing approach in 1910, when he sent out sales teams with horse-drawn wagons loaded with phonographs and records to canvass customers door-to-door. The seller offered to leave a machine and some records in the home for a free trial, then returned in a few days to complete the sale or remove the phonograph. The wagon operators sent detailed reports to West Orange—information the company used to plan further marketing schemes.

The experience of Roy Matton, a traveler for the Silverstone Talking Machine Co. in St. Louis, is typical. On October 15, 1910, Matton visited forty-one homes. He placed four phonographs on free trial, sold a Fireside machine left on a previous call, and sold no phonographs on the first call. The woman at the second stop "would not consider free trial as she claimed she was afraid the children would make her buy same." At the next call, Matton reported, "The lady peeped out of the door then slammed it shut." On his sixth stop, "a German lady" was "out of temper."

Edison dealers were encouraged to stage showroom demonstrations or recitals for the public. Customers could attend Turn Table Comparisons, where they would compare the

Diamond Disc phonograph with competing talking machines. To demonstrate the quality of Edison records, the company sponsored a series of Tone Tests, which featured comparisons between live Edison recording artists and their records played on Edison machines. The goal of the Tone Tests was to demonstrate that Edison records were so precise and realistic that they could not be distinguished from the live performance. Edison recording artist Marie Rappold held one of the first Tone Tests in April 1916 at Carnegie Hall in New York City, and hundreds of similar tests were conducted during the First World War.

In the 1920s, Edison tested new marketing plans to increase record sales. One plan, launched in September 1922, called for the formation of Edison Home Service Clubs—similar to the Book of the Month Club. Each month, subscribers received twenty records from the Edison catalog. They had two days to listen to the recordings and order the ones they desired from a list before mailing the records to another club member.

"DEAFNESS HAS ITS ADVANTAGES. MY OWN DEAFNESS ENABLES ME TO CONCENTRATE MY THOUGHTS AS I'D NEVER BE ABLE TO DO IF DISTRACTED BY NOISE AND CONVERSATION."

The first clubs were organized in Summit and Morristown, New Jersey. Edison based this plan on the belief that he could sell more records through home demonstrations than through mass-circulation advertising. According to company vice president Arthur Walsh, "Edison feels that an actual demonstration of the Edison in the home is the one way to convince people of the Edison's superior tonal quality."

The Sub-Dealer Plan involved placing phonographs in barbershops, ice cream parlors, and other stores. The proprietors of these stores were asked to demonstrate the phonograph and send the names of potential buyers to the local Edison dealer. These marketing plans seemed to work when tried on a limited experimental basis, but to be effective against Victor and other competitors, they had to be implemented on a national scale. Edison, unfortunately, did not have the resources to organize and manage large-scale canvassing plans.

NEXT PAGE: The West Orange lab music room, as it appears today. Experimenters developed recording techniques here in the 1890s. In the early twentieth century, Edison used this room to evaluate music and artists for his record catalog.

TRADE
Thomas A. Edison
MARK

LEFT: A truckload of Edison disc phonographs en route to Rooney & Co., Brooklyn, New York. Edison relied on a national network of wholesalers and retailers to market his phonographs. RIGHT: In 1910, Edison dealers experiemented with horse-drawn wagons to deliver phonographs and records for free home trials.

Edison's dealers complained about the lack of advertising. In July 1923, C. A. Lloyd wrote the company, "It has been a long time since we have seen an Edison advertisement in any of our national periodicals that I have forgotten what they look like. . . . There is no danger of anyone forgetting what a Victor advertisement looked like or a Brunswick or a Columbia advertisement for the reason that you can scarcely pick up any of our best magazines but the very beautiful, colored, artistic advertisements of these companies stare you in the face." The company's response summarized Edison's view of advertising: "It is his belief and he has proved that Edison phonographs placed in the homes for demonstration sell to a very great extent. Mr. Edison argues that the superior tonal quality of the Edison when heard far outweighs the printed advertising of our strongest competitor."

DURING THE EARLY 1920S, phonograph manufacturers faced competition from radio. The first commercial radio station, Pittsburgh's KDKA, began broadcasting in 1920. By 1922 there were thirty commercial radio stations in the United States. This number climbed to 556 in 1923. Hobbyists with a little knowledge and some radio parts could cobble together their own receivers, but manufacturers began producing radios under the patents of the Radio Corporation of America (RCA). Radio production increased from 100,000 sets in 1922 to 500,000 in 1923. By that year, 400,000 American households had radios.

Consumers began trading in their phonographs for radios, and a number of phonograph dealers began selling radios to regain lost business. Unlike Victor and other phonograph producers, who began manufacturing combination radio-phonograph sets, Edison refused to enter the radio business. He did not oppose radio—in time, he believed it would become an important source of news and information—but because the medium distorted the sound of his recordings, he did not believe it would be a popular source of entertainment. Edison also did not think the radio business would be profitable for his phonograph dealers because the field of licensed radio manufacturers was crowded, and any profits dealers made on radio sales would be eroded by service and repair calls.

In 1928, the company overcame Edison's opposition to radio and purchased the Newark, New Jersey–based Splitdorf Radio Co., which had a license to manufacture radios under RCA's patents. Edison's youngest son, Theodore, took charge of the design of an Edison combination radio-phonograph, but it was too late to save the phonograph business. As Edison predicted, the radio business was not sustainable in the late 1920s; there were too many radio manufacturers who began to cut prices to increase sales. The Edison Co. could not compete in this market.

Edison never regained his early leadership in the phonograph industry. In 1919 he produced 7.2 percent of the phonographs and 11.3 percent of the records manufactured in the United States. By the mid-1920s he controlled 2 percent of the domestic record market. As one Edison dealer lamented in 1926, "Instead of Edison being the most popular machine it is one of the least popular. The Edison, the original, thew Daddy of them all should not take second place for any of them." After years of steady losses, Edison stopped producing entertainment phonographs and records in the fall of 1929.

Edison exhibit at the Radio World's Fair, September 1930, Madison Square Garden, New York City.

7

MOTION PICTURES

"I AM EXPERIMENTING UPON AN INSTRUMENT WHICH DOES FOR THE EYE WHAT THE PHONOGRAPH DOES FOR THE EAR WHICH IS THE RECORDING AND REPRODUCTION OF THINGS IN MOTION."

DURING THE EARLY 1890S the West Orange laboratory designed a motion picture camera called the kinetograph and a peephole film viewer called the kinetoscope. Even more significantly, the lab constructed one of the world's first motion picture studios and launched a kinetoscope exhibition business that became the basis of the modern motion picture industry.

Other inventors had worked on motion picture cameras in the late 1880s, including the English photographer William Friese-Greene, who patented a camera in June 1889; English lawyer Wordsworth Donisthorpe, who patented his own camera in August 1889; and Louis Aimé Augustin Le Prince, who developed a multiple-lens camera. Unlike Edison, none of these inventors had the resources to commercialize motion picture technology, however.

In contrast to most of his other major inventions, Edison did not have a clear market conception for motion pictures at the beginning of this project. Edison was a successful innovator because he combined his technical problem-solving capabilities with a sophisticated

PAGES 120-121: Rehearsal of a scene from an unidentified motion picture at Edison's Bronx, New York, studio. ABOVE: Edison's Home Projecting Kinetoscope. Initially ambivalent about the market for motion pictures, Edison came to appreciate its educational value. "I had some glowing dreams about what the camera could be made to do and ought to do in teaching the world things it needed to know."

understanding of the manufacturing and marketing requirements of his inventions. For example, he researched and understood the market for the phonograph and, later, the storage battery, Portland cement, and rubber—knowledge that helped shape technical research and manufacturing and marketing strategies for these endeavors.

When he began thinking about the possibilities of motion pictures, however, he did not articulate a vision of how this technology might be marketed and used. Eventually, the laboratory introduced a commercial motion picture camera and began producing cameras, projectors, and films. The business became one of his most profitable ventures, but Edison was never as actively involved in the motion picture business as he was in other inventions.

The reasons for this are unclear, but it may stem from the context in which Edison approached innovation at West Orange in the late 1880s. Edison expected the laboratory to turn out a variety of small-scale inventions that he would rapidly manufacture and market. Accordingly, he drafted lists of projects that his staff could work on when they were not busy with more important research, like the phonograph and ore milling. Occasionally, Edison would give an idea to an experimenter to see how it could be developed. Many of these ideas, such as artificial silk or a snow-removal machine, went nowhere. In the case of motion pictures, Edison assigned the project to William K. L. Dickson, an experimenter who had considerably more success.

During the 1880s photographer Eadweard Muybridge conducted human and animal motion studies at the University of Pennsylvania.

WHAT BECAME A PROVIDER of mass entertainment in the early twentieth century began as an attempt to combine the reproduction of sound with moving pictures. Eadweard Muybridge, an English photographer who had produced a series of images to study animal motion in the late 1870s, influenced

Edison's interest in this subject. Muybridge had also invented the zoopraxiscope, a device that used a rotating glass wheel and a slotted disc to project pictures. The zoopraxiscope was based on an optical phenomenon called the "persistence of vision," in which a series of images viewed in quick succession create the illusion of movement.

Muybridge visited West Orange on February 27, 1888, to talk with Edison about combining the zoopraxiscope with the phonograph. Edison was intrigued by the idea, but he did not give it much thought until later that year, in October, when he drafted a patent caveat in which he proposed a system of motion pictures: a device to record the images, a device for viewing them, and an instrument that merged viewing pictures and listening to sound in the same experience.

Edison envisioned a camera that would produce a large number of images in quick succession. Muybridge had used multiple cameras to record a relatively small number of images (between twelve and twenty-four in each series) to study animal motion. Edison, in contrast, wanted a camera that would take numerous small images in sequence—at least eight pictures per second, preferably twenty-five. (Edison later increased the film rate to forty-five images per second, but when the motion picture industry introduced sound films in the late 1920s, it adopted twenty-four frames, the

TOP: Muybridge's zoopraxiscope influenced Edison's interest in motion pictures. BOTTOM: Muybridge's photographs of a horse in motion. California industrialist Leland Stanford hired Muybridge in 1872 to determine if all four legs of a galloping horse were off the ground at the same time.

standard still used today). This camera would produce thousands of tiny images (up to 42,000) that would be pasted in strips to a metal cylinder. The images could then be viewed through a microscope as the cylinder rotated.

To achieve his original goal of combining moving pictures with a sound recording, Edison attached an image cylinder and a standard phonograph cylinder to the same shaft. For persistence of vision to work, however, each image had to stop for a fraction of a second to allow the eye to register the picture. While the image cylinder had to turn intermittently, the phonograph recording had to rotate at a steady, even rate. From his work on stock tickers, Edison could design machines that moved paper tape intermittently, but achieving intermittent with continuous motion in the same device was more difficult, and the West Orange laboratory never managed to synchronize moving images with sound recordings on the same machine.

Edison directed William K. L. Dickson to take charge of motion picture research in June 1889. Dickson was an aspiring inventor of Scotch-American descent who was born in France in 1860. He had come to the United States in 1879, shortly after sending an unsuccessful job application to Menlo Park. In 1883, Dickson began working in the dynamo testing department of the Edison Machine Works. He joined Edison's staff at West Orange in 1887 to supervise the metallurgical laboratory and also served as the lab's official photographer.

William K. L. Dickson was Edison's principal motion picture experimenter during the early 1890s. He also worked on the ore milling project and served as the West Orange lab's official photographer.

Dickson began by improving Edison's cylinder approach. The small cylinder photographs were difficult to view, and the cylinder's curvature produced an unfocused image. He increased the size of the images to one-quarter of an inch and attempted to apply a photographic emulsion directly to the cylinder. During this early research, Dickson produced a series of short films, collectively known as *Monkeyshines*, of a lab employee gesturing broadly against a black background.

The development of motion pictures required Dickson and other experimenters to solve a number of technical problems. The most significant challenges were designing a camera and film medium to record a large number of images. In the 1880s, photographers produced pictures on glass plate negatives, but it would be difficult to make a large number of images on glass without breaking the negatives. The development of celluloid and paper-based photographic film provided a solution to this problem.

Celluloid was a plastic material made out of cellulose nitrate, a compound first developed in the 1850s. It was commonly used to make such objects as billiard balls before John Carbutt, an English photographer, produced thin sheets of celluloid coated with a photographic emulsion during the 1880s. Dickson began experimenting with celluloid film, which could be cut into strips of any size.

Early in his career, French scientist Étienne-Jules Marey invented instruments to study human blood circulation. In the 1880s he designed a camera to photograph birds in flight.

In 1888, inventor George Eastman introduced his inexpensive Kodak camera, which used photographic film on rolls of paper. The film was preloaded at the factory. After taking their pictures, consumers sent the camera back to Eastman's factory in Rochester, New York, where the film was developed and prints sent back to the customer. The development of paper-based photographic film revolutionized photography and changed the direction of Edison's motion picture research.

In August 1889 Edison traveled to Paris to attend the Exhibition Universelle. In Paris he met Étienne-Jules Marey, a French physiologist who had designed a camera capable of taking continuous exposures on a strip of paper-based film. Marey's camera convinced Edison that the lab was on the right track, and when he returned to West Orange in October, he drafted a new patent caveat that described what we recognize as a modern motion picture camera. "The sensitive film is in the form of a long band passing from one reel to another in front of a square slot . . . on each side of the band are rows of holes exactly opposite each other & into which double toothed wheels pass." These holes, Edison claimed, would advance the film through the camera at the rate of ten images per second.

Because motion pictures were not a priority for Edison, Dickson did not resume research until October 1890, when Edison assigned William Heise, an experimenter with printing telegraph experience, to help him. Dickson worked on the optical components of the camera and the film, while Heise designed the mechanical features that moved the film through the camera.

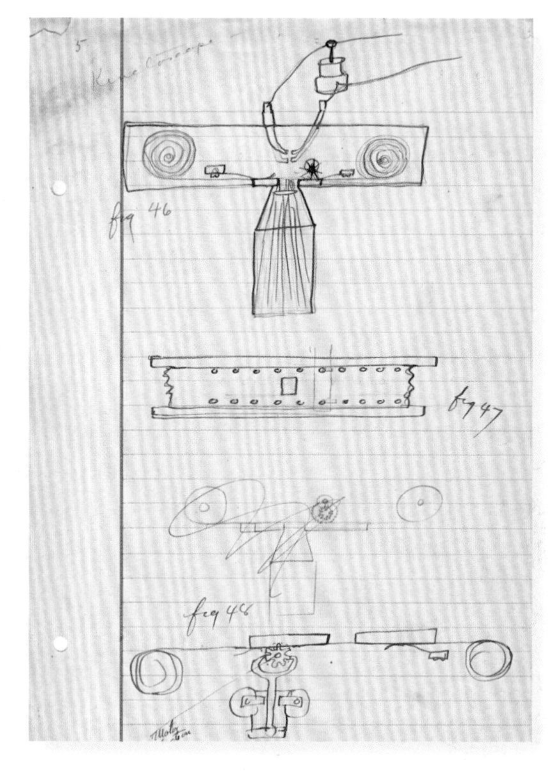

Edison's December 1889 drawing, prepared for a patent caveat, illustrates a significant change in his conception of motion pictures, from cylinder-based images to photographs on strips of film.

By the spring of 1891, they had designed a horizontal-feed camera called the kinetograph, which exposed images on strips of perforated film three-quarters of an inch wide. The camera had a shutter and escapement mechanism that allowed it to stop the film for a fraction of second, make the exposure, and then advance the strip for the next exposure. Escapement mechanisms, commonly used in watches, are gears and levers that regulate intermittent motion. Edison used escapements in his stock ticker and other telegraph inventions—experience that informed his design of the motion picture camera. As the *Literary Digest* reported in 1894, "The speed of this photographic work is certainly astonishing. Forty-six impressions are taken each second, which is 2,760 a minute and 165,600 an hour." Dickson used this camera to produce several short experimental films, including a lab worker smoking a pipe and another swinging a set of Indian clubs.

To view films, the laboratory designed the kinetoscope, a wooden box with an eyepiece at the top. Compartments inside contained an electric lamp, a battery-powered motor, and a fifty-foot ribbon of positive celluloid film arranged on a series of rollers and pulleys. The motor moved the film at the rate of forty-six frames per second between the eyepiece and the electric lamp. On May 21, 1891, Dickson exhibited a film of him tipping his hat on a makeshift kinetoscope to a group representing the American Federation of Women's Clubs, which visited the lab after a luncheon at Glenmont.

Edison submitted patent applications for the kinetograph and kinetoscope on August 24, 1891. In the fall of 1891, Dickson and Heise continued improving the kinetograph, focusing on designing a vertical-feed camera. Because of this ongoing research, Edison was not ready to consider marketing motion pictures until the end of 1892.

EDISON WAS AMBIVALENT about the market for motion pictures. In May 1891 he told the *Chicago Evening Post*, "This invention will not have any particular commercial value. It will be rather of a sentimental worth." In that same interview, he foresaw what would become the predominant form of home entertainment in the twentieth century. "When this invention shall have been perfected a man will be able to sit in his library at home and, having electrical connection with the theater, see reproduced on his wall or a piece of canvas the actors and

The West Orange lab's first motion picture camera, the strip kinetograph. Dickson later recalled the first subject he filmed in early 1889: a troupe of dancing bears.

The Black Maria, considered to be the world's first motion picture studio, was one of a number of temporary buildings constructed at West Orange. In the 1920s, years after it was dismantled, Edison remembered, "It was a ghastly proposition for a stranger daring enough to brave its mysteries—especially when it began to turn like a ship in a gale."

hear anything they say." In February 1894, he wrote Muybridge, "I am doubtful there is any commercial feature in it & fear that they will not earn their cost—these zootropic [*sic*] devices are of too sentimental a character to get the public to invest in."

He may have thought that the novelty of motion pictures, like that of his tinfoil phonograph in 1878, would soon fade. Nevertheless, he gave his secretary, Alfred O. Tate, permission to arrange a kinetoscope exhibit at the World's Columbian Exposition in Chicago, scheduled to open in May 1893. To supply Tate with machines, Edison gave one of his machinists, James Egan, a contract to make twenty-five kinetoscopes.

Workers began constructing a motion picture studio on the grounds of the laboratory in December 1892. Located near Building 4, the studio was a fifty-by-eighteen-foot wood building with a twenty-one-foot-high pitched roof, built on a graphite pivot that allowed the staff to turn the studio on a wood track. According to the *New York Sun*, "Half of the roof

Edison peephole kinetoscope, 1895. The viewer's ear tubes reveal that this machine was equipped with a phonograph to allow synchronized sound.

could be raised or lowered like a drawbridge by means of ropes, pulleys and weights, so that the sunlight could strike squarely on the space before the machine." Because electric lights were not bright enough to expose motion picture film, filmmakers took advantage of the best available natural light. Edison's staff nicknamed the studio the Black Maria because it resembled the nineteenth-century police wagon used to transport prisoners. "With its great flapping sail-like roof and ebon complexion," Dickson wrote, the Black Maria "has a weird and semi-nautical appearance."

Inside, performers worked in front of the camera in a large room with black tar paper–covered walls. The camera lens had a fixed focal length and could not be adjusted—zoom lenses first became available in 1902—so the operator moved the camera back and forth on a table mounted on a track to film close-up scenes. At one end of the building, the windows of a small room were shaded red in order to prevent exposure of the film before it was loaded into the camera.

Among the earliest films made in the Black Maria were *Blacksmith Scene*, photographed sometime in April or early May 1893 and featuring Edison employees Charles Kayser and John Ott, and *Horse Shoeing*, depicting Dickson and a blacksmith shoeing a horse. Both of these films were exhibited at the first public demonstration of the kinetoscope at the Brooklyn Institute of Arts and Sciences on May 9, 1893. In January 1894, Dickson filmed Edison experimenter Fred Ott in what became the first copyrighted motion picture, officially known as *Edison's Kinetoscopic Record of a Sneeze*.

James Egan did not complete his manufacturing contract until March 1894. Tate missed his opportunity to demonstrate the kinetoscope in Chicago, but working with several business partners, he planned to open a kinetoscope exhibition in New York City. In a former

Dickson photographed *Fred Ott's Sneeze* in the Black Maria on January 7, 1894. The short 45-frame film is the first copyrighted motion picture.

"IF THE MOTION PICTURE HAS DONE NOTHING ELSE, IT HAS BEEN THE GREATEST QUICKENER OF BRAIN ACTION THAT WE HAVE EVER HAD."

Peter Bacigalupi's kinetoscope parlor at 946 Market Street, San Francisco. Bacigalupi later became one of the largest wholesalers of Edison phonographs on the U.S. West Coast.

shoe store at 1155 Broadway, Tate installed ten kinetoscopes equipped with coin-slot attachments, arranged in two rows of five each. Patrons would be charged twenty-five cents ($6.74 today) to see the films in one row.

The exhibit opened on Saturday, April 14. Tate recalled later that the kinetoscope parlor was supposed to open on the following Monday, but he decided to let the public in early to raise enough money for a pricey meal at Delmonico's. He hoped to close at six that night, but the crowds kept the exhibit busy until late into the evening. When Tate and his brother finally locked the door at one in the morning, they left with $120.

Another kinetoscope parlor opened in Chicago in May, and five kinetoscopes were installed in a San Francisco phonograph parlor in early June. Later that summer, kinetoscope exhibits opened in other cities, including Atlantic City, Boston, and Washington, D.C. The first international kinetoscope parlor opened in London in October.

In August 1894, Edison gave Norman Raff and Frank Gammon—two entrepreneurs who had been involved in the phonograph business—the exclusive right to sell kinetoscopes in the United States and Canada. They agreed to purchase ten kinetoscopes per week for $200 ($5,390 today) each. In turn, they sold the machines to exhibitors for as much as $250 ($6,910 today). Raff and Gammon also paid Edison nine dollars ($249 today) for each kinetoscope film.

Dickson ramped up film production in the Black Maria in 1894, bringing a cavalcade of acrobats, dancers, boxers, vaudeville artists, circus acts, and trained animals to perform in front of the camera. In July, Professor Henry Welton brought his boxing cats and wrestling dog to the Black Maria. Annabelle Whitford performed her *Butterfly Dance* for the camera in August.

On September 7, Dickson filmed a six-round boxing match between former world champion James J. Corbett and Peter Courtney, a heavyweight from Trenton, New Jersey. Corbett knocked out Courtney in the sixth round. Later that month, Buffalo Bill brought a troupe of Native American dancers—including Parts His Hair, Charging Crow, Dull Knife, and Crazy Bull—from his Wild West show to perform their Buffalo Dance in full war paint. On October 17, Dickson filmed animal trainer Ivan Tschernoff's dog, Leo, performing somersaults. Annie Oakley, a star performer in Buffalo Bill's Wild West Show, came to West Orange in November to film a rifle-shooting demonstration. In 1894 Dickson and his camera operator produced seventy-five films in the Black Maria studio.

Edison at the controls of a motion picture projector in the West Orange library.

DURING ITS FIRST YEAR, the kinetoscope business was highly profitable. From April 1894 to February 1895, the Edison Manufacturing Co. produced $177,847 ($4.91 million today) worth of kinetoscopes and films. As Edison anticipated, the kinetoscope's novelty soon faded, and receipts began to decline. For the twelve months beginning in May 1895, motion picture sales had dropped to $49,896 ($1.38 million today).

In April 1895, Edison responded to declining kinetoscope sales by introducing the kinetophone—a kinetoscope attached to

Edison Vitascope, April 1896.

a phonograph that allowed customers to view a film while listening to a cylinder recording through ear tubes. In one experimental film produced for the kinetophone, Dickson danced with another lab employee as a third worker played a violin. A phonograph captured the music as the camera recorded the images. Despite this innovation, the kinetophone was not a commercial success.

Raff and Gammon believed that kinetoscope patrons wanted more interesting films, so in the summer of 1895 they directed an employee, Alfred Clark, to film a series of historical reenactments. Clark produced *The Burning of Joan of Arc* and *The Rescue of Captain John Smith.* In *The Execution of Mary, Queen of Scots,* Clark used stop-action photography to depict the beheading of Queen Mary (portrayed by Edison lab employee Robert Thomae). As the executioner's ax fell on Thomae, Clark stopped the camera and replaced the actor with a dummy. The two shots were later spliced together to appear as a continuous scene.

These films, however, failed to revive the kinetoscope business, and Raff and Gammon were considering leaving the motion picture business when they saw the phantascope, a projector invented by C. Francis Jenkins and Thomas Armat in December 1895. The next month Raff and Gammon asked Edison to manufacture the projector under his own name. Edison agreed, and the Edison Manufacturing Co. began producing a projector called the Vitascope. Koster & Bial's Music Hall on 34th Street (near Herald Square) hosted the first commercial exhibition of the Vitascope on April 23, 1896.

THE VITASCOPE WAS NOT SUCCESSFUL. Edison ultimately manufactured only seventy-three Vitascopes, and in 1897 he introduced his own projector. Nevertheless, film projection was an important technological advance in the motion picture business. In contrast to the fifty-foot kinetoscope films, projectors allowed the production of longer movies.

In another technical advance, Edison introduced a portable motion picture camera in May 1896 that allowed film production to move out of the Black Maria. In that same month, camera operator James White began filming scenes in New York City. He photographed a Central Park fountain and a street scene at Herald Square on May 11. Later that month, he took his camera to Scranton, Pennsylvania, to film a Knights of Templar parade. In early June he shot several films of Niagara Falls and amusement attractions at Coney Island, and on March 4, 1897, he filmed President William McKinley's first inauguration in

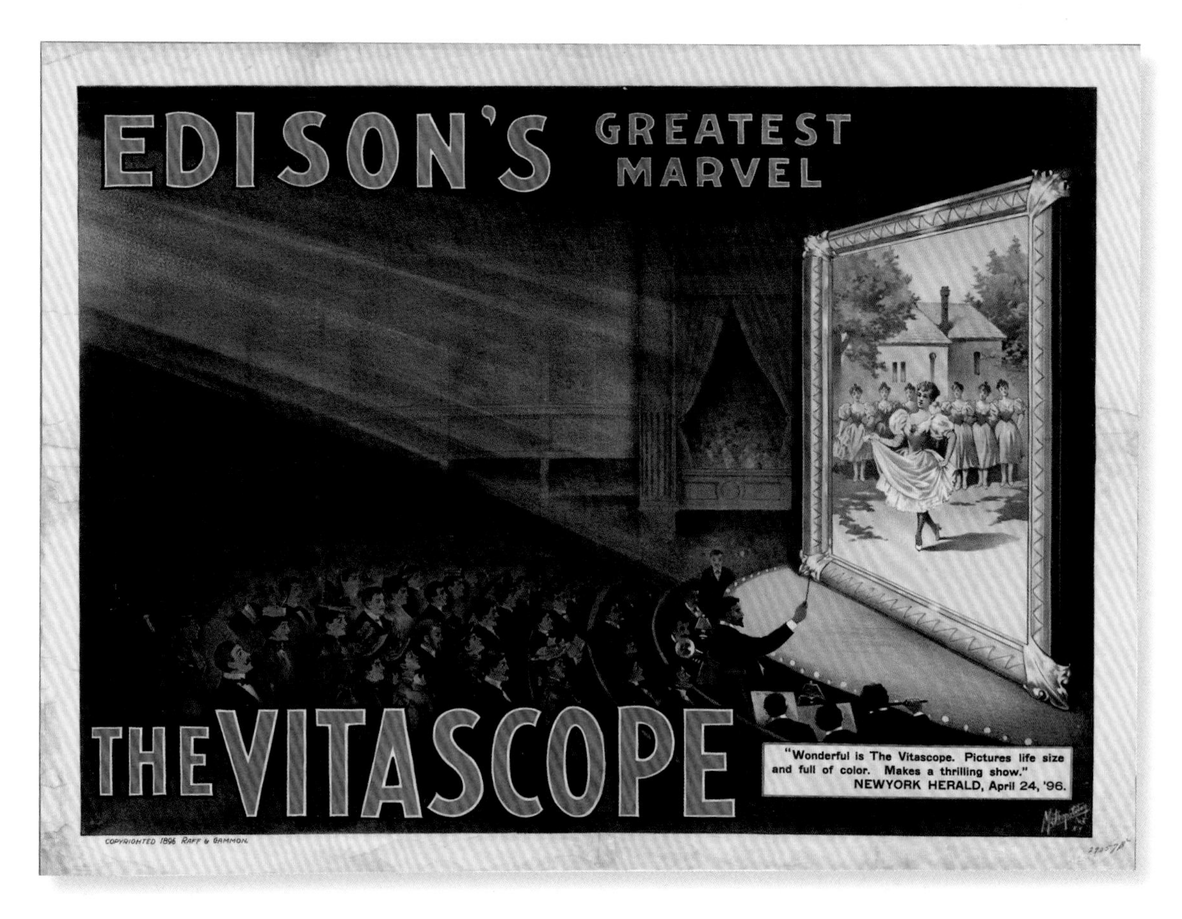

The introduction of film projectors in 1896 changed movie exhibition from a solitary experience to a communal activity.

Experimenter Charles H. Kayser works on a motion picture projector in the West Orange lab, ca. 1897.

Washington, D.C. White also took his camera to Yellowstone National Park, Colorado, and Mexico in 1897, followed by China, Japan, and the Hawaiian Islands in 1898.

In the fall of 1900, the Edison Manufacturing Co. opened a new studio on the top floor of a building at 41 East 21st Street in Manhattan. The new studio improved the company's access to stage and vaudeville performers. In 1907, the company opened a larger studio in the Bronx, New York. The Black Maria, no longer needed, was torn down in 1903. A full-scale replica of the Black Maria, constructed in 1954, stands today at Edison's West Orange laboratory.

In 1900, Edwin S. Porter, a projectionist who began working for Raff and Gammon in 1896, became Edison's chief cameraman. Porter shifted film production from the "actualities" of the late 1890s to narratives, or movies that told stories. One of Porter's earliest films was *Kansas Saloon Smashers*, a burlesque of temperance reformers. Another 1901 production, *Terrible Teddy, the Grizzly King*, poked fun at Vice President Theodore Roosevelt's hunting exploits.

LEFT: Film production at Edison's Bronx motion picture studio. RIGHT: Actors at the Bronx studio rehearse a scene from *Hope, A Red Cross Seal Story*, a drama released on November 16, 1912.

In November 1902, Porter enlisted the aid of the East Orange, New Jersey, Fire Department to film *Life of an American Fireman*, which focused on the rescue of a woman and child from a burning building. Other scenes showed the firefighters responding to the alarm, sliding down poles, hitching their horses to a fire engine, and dashing to the fire.

One of Porter's most successful films was *The Great Train Robbery*, based on a play of the same title by Scott Marble. Released in December 1903 and filmed at Edison's Manhattan studio and on location in New Jersey, the movie portrayed the robbery of a train by a gang of horse-mounted bandits and their eventual apprehension. Considered by film critics as the first American "Western," it incorporated several innovative film techniques, including the intercutting of scenes to depict simultaneous action.

THE U.S. PATENT OFFICE did not grant Edison's motion picture patents until August 1897 because some of the features in the applications had been anticipated by other inventors. The lapse allowed several domestic producers, including the International Film Co. and Edward Amet, to enter the motion picture business. Another Edison competitor was the American Mutoscope Co., a firm organized in 1895 by William Dickson (who left Edison in April 1895) and other inventors to develop a projector called the Biograph.

In December 1897, Edison began a series of legal actions to force these competitors out of business. Some rivals, such as the International Film Co., left without a fight; others agreed to license Edison's patent. The American Mutoscope Co., however, fought Edison in protracted litigation that would not be resolved until 1907. Over this ten-year period, Edison's patents were upheld in court and then invalidated on appeal. The reissue in 1902 and 1904 of the patents prompted a new round of litigation with the Mutoscope Co. In 1907, Edison's patents were finally upheld.

To end expensive litigation, on December 18, 1908, Edison and the American Mutoscope Co. agreed to pool their patents in a trust called the Motion Picture Patents Co., which licensed production and distribution of films to the Edison Manufacturing Co. and eight of its competitors. Frank Dyer—

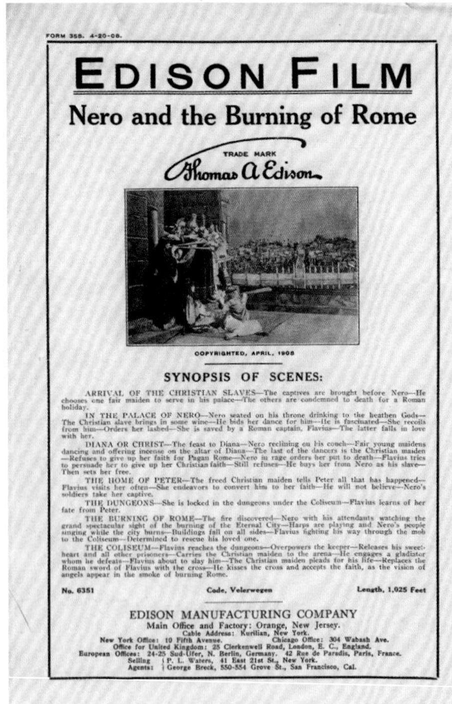

TOP: Still from *The Great Train Robbery* (1903). BOTTOM: Ad for Edwin S. Porter's 1908 film *Nero and the Burning of Rome.* A *Moving Picture World* critic opined, "This subject is spectacular, contains many elements of human interest and possesses the dignity of history."

Edison kinetophone, ca. 1912. This late attempt to combine motion pictures with recorded sound was not a commercial success.

Edison's patent attorney, who played a key role in negotiating the patent pool—explained that the Edison Co. preferred this arrangement because it did not have the facilities to provide the entire domestic market with motion picture films. Film exhibitors were required to obtain licenses and pay royalties to the trust members. The Motion Picture Patents Co., which controlled up to 90 percent of the American film market, helped stabilize the industry.

A number of independent filmmakers, however, refused to participate in the patent pool. In 1912, the U.S. government sued the Motion Picture Patents Co. for violation of the Sherman Anti-Trust Act. On October 1, 1915, U.S. District Court Judge Oliver B. Dickinson ruled that the trust could not use patent rights to create a monopoly and ordered the trust dissolved.

AFTER 1900, the West Orange lab no longer produced significant improvements in motion picture technology. Edison supported unsuccessful research aimed at developing color motion picture film, and in 1910 he received a patent for a device to reduce motion picture flicker.

EDISON HOME PROJECTING KINETOSCOPE

Edison believed that motion pictures could be an effective educational resource. In July 1909 he wrote the Queensland, Australia, commissioner of public health: "Kinetoscope pictures in schools would be an epoch in the common schools—technical subjects & every branch of human endeavor could be shown in actual operation. You couldn't keep the children away, even by the police, knowledge & schools would then be attractive."

To promote this goal, in 1910 Edison began developing a motion picture projector for noncommercial use in schools, homes, churches, and civic organizations. In 1912, he introduced the Edison Home Projecting Kinetoscope, priced between sixty-five and ninety dollars ($1,500 and $2,150 today), depending on the model. The laboratory also reformatted the frame of the commercial motion pictures so that sixteen minutes of film (the standard for a one-reel movie) would fit on a smaller reel.

Edison developed a catalog that included films on various scientific subjects. Eventually, the Home PK catalog featured 160 titles—on subjects ranging from the pink granite industry

Titles in the Edison Home PK film catalog included *Importing Cattle from Mexico into the U.S.*, *Orange Growing*, and *Quaint Spots in Cairo, Egypt.* Edison also sold sets of lantern slides for the Home PK, depicting scenes from—among other places—Ireland, Norway, India, and Venezuela.

Introduced in November 1912, Edison's Home Projecting Kinetoscope was his attempt to market motion pictures for home and school use.

of Milford, Massachusetts, and the South African whaling industry to the Civil War battlefields of Chattanooga, Tennessee—at prices ranging from $2.50 to $20 ($59.80 to $479 today). The initial reaction of educators to the Home PK was favorable, but the machines and films were too expensive for most schools. Thus, in the summer of 1913, Edison decided to stop spending money on the production of new educational films.

The high cost of the projectors and the lack of an extensive film catalog also discouraged residential users. In 1917, Mrs. S. E. Sinclair of Winter Hill, Massachusetts, complained to the company about her inability to get new films:

> We could hardly afford the machine but were anxious to have a motion picture show at home for our boys rather than send them out to the theatres where there is so much objectionable for little ones to see. And so it was necessary to buy on the installment plan, and the hard part about the matter is that when we had finished paying for the machine, they told us we couldn't get any more films for it.

The Edison Home PK ultimately did not succeed in the market. By March 1915, the company had shipped nearly 2,500 machines to dealers. Only 500 were sold. Slightly more than 1,000 machines sat in a warehouse, and none of the approximately 1,200 Home PKs shipped abroad—primarily to England and Germany—were sold.

That same year, he demonstrated his kinetophone, a renewed attempt to introduce talking motion pictures. Essentially a film projector connected by belts and pulleys to a phonograph, the kinetophone debuted at a New York City theater in February 1913. Over the next several months, it was exhibited in cities throughout Europe, Asia, and South America, but poor sound quality prevented its commercial success.

The Edison Co. continued to produce motion pictures up to 1915. The Edison catalog included one- and two-reel dramas, comedies, historical reenactments, and industrial documentaries. Edison's comedies included *Andy Has a Toothache*, the eighth of a series released on July 8, 1914. In this installment, Andy wakes up with a toothache, and mayhem ensues when his mother and dentist attempt to fix the problem. *Christian and Moor* (released August 1, 1911) was an Edison drama that was filmed on location in Havana, Cuba. It depicted a love affair between a Moorish princess and a Christian knight in medieval Spain. *The Life of Abraham Lincoln* was a two-part drama released on February 26, 1915. According to the Edison catalog, "From the scene in front of the log cabin to the assassination at Ford's

J. Searle Dawley directing a scene from *Christian and Moor* at Morro Castle in Havana, Cuba, 1911.

774

Title Christian ang Moor

Photographed by Henry Cronjager

On ——— 1911 at Havana Cuba

Three Bromide prints printed by B S Dawley

On July 5th 1911 at ~~West Orange, N. J.~~ New York N.Y.

One printed copy of title deposited in Post Office at Orange,

N. J., on Aug 4 1911 by J. Bredamus at 6:00 P.M

Addressed Librarian of Congress Washington, D. C.

Receipt 91565

Two Bromide copies of film deposited in Post Office at Orange, N. J.,

on Aug 4 1911 by J. Bredamus at 6:00 P.M

Addressed Librarian of Congress Washington, D. C

Registered package No. 1058

Receipt 91565

Length of film Nine hundred ninety five feet

Sample enclosed compared by W K Green

First film shipped on ... 190...

By ...

To Release Date August 1st 1911

#774

Form 509

Envelope containing copyright registration prints for *Christian and Moor*. To obtain copyright, the Edison Co. submitted positive prints of each scene to the Library of Congress.

Theater in Washington, one is gripped. Nothing has been left undone to make this a consummate review of Lincoln's life."

The Bronx studio also produced fantasies, such as *A Trip to Mars* (released February 18, 1910), a film about a chemistry professor who discovers a way to reverse gravity and travels to Mars, where he encounters strange half-tree, half-human monsters. The studio also filmed adaptations of literary works, including Mark Twain's *The Prince and the Pauper* (released August 13, 1909), Mary Shelley's *Frankenstein* (released March 18, 1910), and Charles Dickens's *Martin Chuzzlewit* (released June 10, 1912). Charles Ogle, the actor who portrayed Dr. Frankenstein's monster, also played George Washington in *The Battle of Bunker Hill* (released August 8, 1911).

The Edison Co. produced films that adhered to proper morals and values. Accordingly, the studio avoided making movies that depicted brutal, salacious, or criminal acts. As Edison Co. manager Frank Dyer wrote in April 1910, "When the works of Dickens and Victor Hugo, the poems of Browning, the plays of Shakespeare and stories from the Bible are used as a basis for moving pictures, no fair-minded man can deny that the art is being developed along the right lines."

Still from an unidentified Edison film.

Before 1911, Edison's motion picture business was one of his most profitable ventures. The annual profit from the sale of films and motion picture equipment from 1908 to 1911 ranged between $200,000 and $230,000 ($5.1 million and $5.8 million today). However, motion pictures sales began to decline in 1911, and by 1914 Edison's share of the business had dropped to 8 percent. For the fiscal year ending March 1, 1916, Edison's motion picture sales were $60,000 ($1.2 million today). For the following fiscal year, ending March 1, 1917, sales had declined to $45,000 ($790,000 today), and for the nine months between March and December 1917, the Edison Co. sold only $15,000 ($263,000 today) worth of motion picture equipment and films. In 1918, Edison sold his studio and film catalog to the Lincoln & Parker Film Co. for $150,000 cash and $200,000 in stock ($2.2 million and $2.9 million today, respectively).

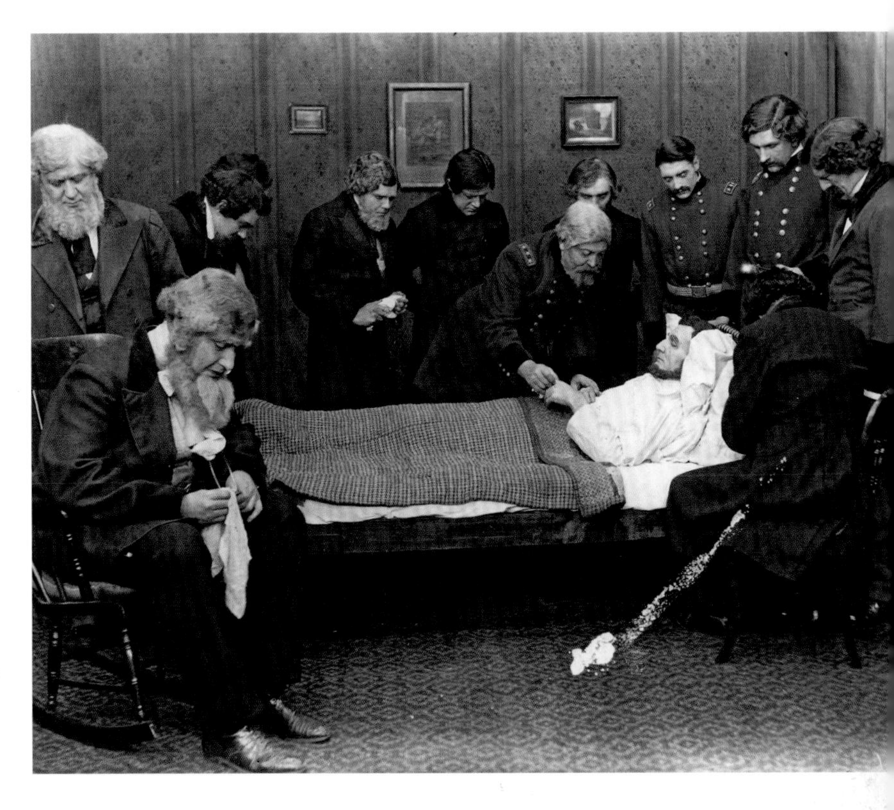

Still from *The Life of Abraham Lincoln* (1915).

A major cause of this decline was Edison's failure to keep up with changing consumer demand for motion pictures. Movie patrons wanted longer feature films, and new independent producers like Adolph Zukor's Famous Player Film Co. (the forerunner of Paramount Pictures) began meeting this demand by producing longer films (four and six reels), in contrast to Edison's one- and two-reel films, which by this time exhibitors considered filler shown between features. The independent film companies also began to film recognizable stars like Mary Pickford, who signed a lucrative contract with the Independent Moving Picture Co. in 1911 and with Adolph Zukor in 1912.

Edison film distributor George Kleine succinctly identified the company's problem in December 1917: "For two years we have been selling films which met with little commercial respect. . . . We have been told that our pictures are good, our methods commendable and that all of us, including particularly the Edison Co., were most likeable and worthy of praise, but our pictures did not draw at the box office. You know, they do not draw people."

32

ORE MILLING

"THERE IS MORE MONEY IN IRON THAN IN GOLD."

EDISON SPENT MUCH OF THE 1890S developing a process for mining low-grade iron ore. His ore milling plant at Ogdensburg, New Jersey—the most expensive and complicated engineering project of his career—was based on a simple idea: Use powerful electromagnets to separate iron ore from rock.

Edison was interested in ore milling as early as May 1875, when he included "a cheap process for the extraction of low grade ores" on a list of proposed experiments. In 1879 he experimented on a process to magnetically separate platinum, which he sought for his electric light filament from the tailings, or waste, of gold mines. He also considered separating gold ore but soon focused on iron.

With the support of two Edison Electric Light Co. investors, James Banker and Robert Cutting, in December 1879 Edison organized the Edison Ore Milling Co. to finance

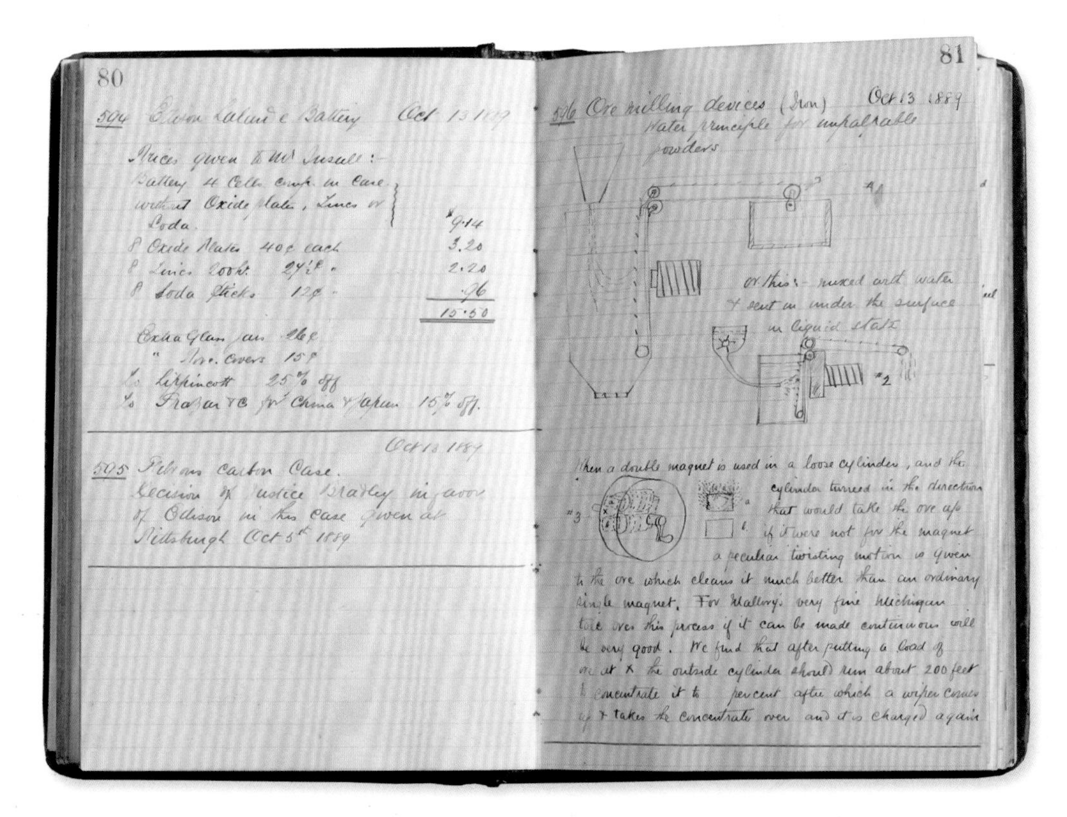

80

504 Edison Lalande Battery Oct 13 1889

Prices given to Mr Insull:—
Battery 4 Cells comp. in Case without Oxide plates, Zincs or Soda. $9.14
8 Oxide Plates 40¢ each 3.20
8 Zincs Compl. 27½¢ " 2.20
8 Soda Sticks 12¢ " .96
15.50

Extra Glass Jars 16¢
" Porc. Covers 15¢
To Lippincott 25% off
To Frazar & Co for China & Japan 15% off.

Oct 13 1889
505 Filament Carbon Case.
Decision of Justice Bradley in favor of Edison in this case given at Pittsburgh Oct 5th 1889

81

506 Ore milling devices (Iron) Oct 13 1889
Water principle for impalpable powders

Or this:- mixed with water & sent in under the surface in liquid state

#2

When a double magnet is used in a loose cylinder, and the cylinder turned in the direction that would take the ore up if it were not for the magnet a peculiar twisting motion is given to the ore which cleans it much better than an ordinary single magnet. For Mallory's very fine Michigan tile ores this process if it can be made continuous will be very good. We find that after putting a load of ore at X the outside cylinder should run about 200 feet to concentrate it to percent after which a wiper comes up & takes the concentrate over and it is charged again

#3

PAGES 146–147: Part of Edison's elaborate ore-handling system at Ogdensburg, the 466 Bee-Line conveyor moved crushed and separated iron ore from stockhouse #2 to the mixing tower, where it was combined with resin and oil to form briquettes. ABOVE: Charles Batchelor recorded these notes on ore milling magnet experiments on October 13, 1889.

ore-processing research. On April 3, 1880, he signed a patent application for a magnetic ore separator, which consisted of a hopper that dropped a stream of crushed iron ore past an electromagnet. The magnet attracted the iron ore away from the nonferrous rock, forcing it to fall into a separate bin.

Edison planned to separate a deposit of black sand on a Long Island beach, but a storm washed it away before he could test the separator. Edison found another deposit of black sand on a beach at Quonochontaug, Rhode Island, but a fire in September 1881 destroyed the separator before it began operating. Edison quickly rebuilt the separator, which processed up to seven tons of black sand per day. The Poughkeepsie Iron and Steel Co., the only iron producer with a furnace that could smelt the ore's fine particles, ordered 200 tons per month. Because of financial problems, however, the company canceled its order before the first shipment left Rhode Island. Edison closed the Rhode Island separator in December 1882.

When he resumed separator experiments in the spring of 1887, Edison increased the capital of the Edison Ore Milling Co. from $350,000 to $2 million in October 1887 to finance the endeavor. On December 27, 1888, he organized the New Jersey and Pennsylvania Concentrating Works to "mine, separate, smelt or otherwise heat, concentrate and deal in iron and other metallic ores."

Although Edison was initially interested in using his ore milling process to separate gold ore, eastern U.S. iron manufacturers encouraged him to apply the technology to iron. During the 1880s, iron and steel production began shifting from blast furnaces in the East to mills in Illinois, Ohio, and western Pennsylvania, which were closer to newly discovered deposits of high-grade iron ore in the Midwest. Midwestern iron ore was easier and less expensive to mine because it was deposited near the surface. It was also low in phosphorus, which makes iron brittle. Many blast furnaces in Pennsylvania's Lehigh Valley closed because they could not afford the cost of shipping iron ore from the Midwest.

There were ample deposits of low-grade iron ore in the East, but it contained a high percentage of nonferrous rock that had to be removed before it could be smelted. Conventional wisdom in the mining industry held that processing iron ore at the mine was expensive and inefficient. Edison turned this view around, betting that he could revive the eastern iron industry by inventing machinery to economically mine and concentrate low-grade iron ore.

Other inventors also worked on developing ore separators in the 1880s. By 1890 there were more than twenty experimental ore separators in the East, most operating on a small scale. Edison planned to process iron ore on a large scale, constructing plants that would mill

The Ogden mill separator building. The three towers housed ore dryers. In 1894 Edison wrote, "The mill as it stands today is the largest crushing plant in the world."

up to 5,000 tons of ore per day. Further, he believed that lower shipping costs—he estimated that he could ship ore to eastern foundries for sixty-eight cents per ton compared to three dollars per ton of ore from the Midwest—would give him a competitive advantage.

In January 1889 Edison established the Edison Iron Concentrating Co., which constructed an experimental ore separator in Humboldt, Michigan. Walter Mallory, a Chicago iron manufacturer who later worked for Edison at Ogdensburg in the 1890s, managed the Humboldt plant, where he encountered a number of technical problems. Dust clogged the machinery, and iron particles stuck to the magnets, so he temporarily closed the mill while Edison worked on solving the problems at West Orange. Mallory reopened the separator in the spring of 1889, but the Edison Iron Concentrating Co. lacked capital for expansion. Edison abandoned the Humboldt project after a fire burned the separator in December 1890.

In the spring of 1889 at the Gilbert Mine in Bechtelsville, Pennsylvania, Edison constructed his first eastern iron ore separator. The endeavor was short-lived; although he was pleased with the performance of this separator, the poor quality of the mine's iron ore forced him to close it in April 1890.

Humboldt and Bechtelsville were small experimental plants that gave Edison and his staff valuable technical experience, but for his process to succeed economically, he needed to test it on a large scale. To find a site for this larger plant, Edison identified iron mines on geological and topographic maps, then sent out teams of surveyors with magnetic dipping needles to survey and map the ore deposits. By 1893 Edison had mapped the primary iron ore deposits from Connecticut to Virginia. "It is certain," he boasted, "that there is no magnetic deposit of ore of any considerable extent in the East that is not known to us."

An engineer stands next to a Corliss steam engine in the Power House. Steam from the large engine powered the crushing rolls and generated electricity for the separator magnets.

Edison found a promising location for his ore milling plant in the highlands of northwestern New Jersey, two miles south of the village of Ogdensburg. The Ogden Mine, which had been worked since 1772, contained a deposit of iron ore three miles long, 600 feet deep, and an average width of 400 feet. It also was near the New York, Susquehanna, and Western Railway, which allowed Edison to easily transport supplies and equipment for the plant, as well as providing rail access to eastern iron foundries. In July 1889, the New Jersey & Pennsylvania Concentrating Works approved the construction of an ore milling plant at the location and, to finance it, increased its capital stock from $30,000 to $150,000 ($757,000 to $3.8 million today). Edison purchased or leased mineral rights for 19,000 acres at Ogden in October.

“The giant rolls are what might be called the spectacular feature of the whole plant . . . to see them seize a 5-ton rock and crunch it with less show of effort than a dog in crunching a bone gives one a vivid sense of the meaning of momentum.” *Scientific American*, January 22, 1898.

Completed in 1890, the Ogden ore milling plant used jaw-type crushers to process iron ore, shipping 100 tons of ore to the Bethlehem Iron Works and the Pennsylvania Steel Co. in April 1891. The Bethlehem Works ordered 100,000 tons in May, but the iron companies complained about the ore's high phosphorus content. Edison also faced several technical problems that prevented the mill from operating at full capacity, including separators that left too much iron in the tailings and separator screens that became clogged with wet or damp ore. More significantly, the plant could not concentrate iron ore at a price competitive with high-grade midwestern ore because of high operating expenses—particularly the labor necessary to mine and load iron ore into the separator. Edison decided that the jaw-type crushers were unsatisfactory and began experimenting on roll crushers. Charles Batchelor, semiretired but an occasional helper of Edison at West Orange, researched the phosphorus problem and found that it was caused by dust clinging to wet crushed ore. To solve the problem, he added a drying and washing step to the ore milling process.

Edison left the operation of the Ogden mill to Harry Livor, the general manager of the New Jersey and Pennsylvania Concentrating Works, and P. F. Gildea, the Ogden plant superintendent. In the spring of 1891, though, Livor resigned following allegations of mismanagement of the company, and Edison fired Gildea after he recommended that Edison abandon the separator and extract ore in the traditional manner.

In June 1891, Edison assumed personal responsibility for the ore milling plant and went to Ogden to "fix all the problems." Walter Mallory, who began working at Ogden in early 1892, later testified that Edison had spent $1.4 million ($35.7 million today) on the ore milling project at that point, primarily to purchase the mining property, construct the mill, and pay for experiments. Before Edison ended the ore milling project in 1900, he would spend much more.

IN APRIL 1892, Edison added a bricking plant to the ore milling process. The Ogden mill shipped its ore to iron foundries in fine particle form, so when the iron producers loaded it in furnaces in large quantities, the fine particles blew out. Edison corrected this problem by developing a resin that, when mixed with the crushed ore, allowed it to be molded into briquettes the size of hockey pucks.

At the end of 1892, Edison decided to dismantle the plant and build a new one from the ground up. To pay for the new plant, the New Jersey and Pennsylvania Concentrating Works increased its stock to $1 million. However, a severe economic depression in 1893 dried up sources of capital, and the next two largest stockholders after Edison refused to invest further in the project. Undaunted, Edison decided to support the ore milling plant with his own money.

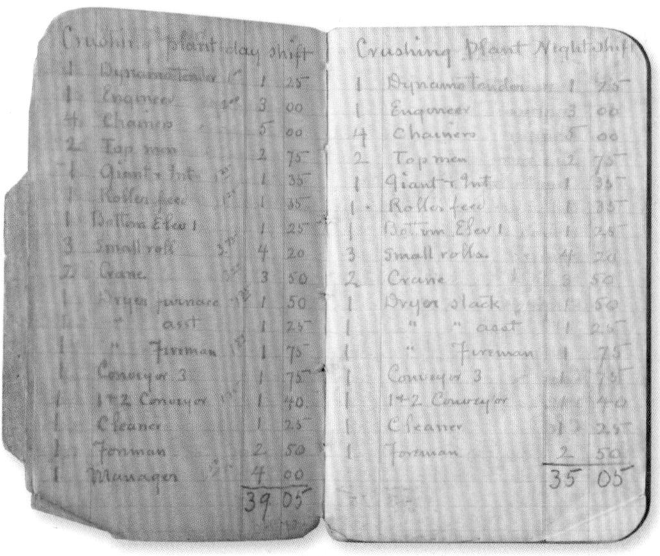

	Crushing plant day shift		
1	Dynamo Tender	1	25
1	Engineer	3	00
4	Chainers	5	00
2	Top men	2	75
1	Giant + Int	1	35
1	Roller feed	1	35
1	Bottom Elev 1	1	25
3	Small rolls	4	20
2	Crane	3	50
1	Dryer furnace	1	50
1	" asst	1	25
1	" Fireman	1	75
1	Conveyor 3	1	75
1	1+2 Conveyor	1	40
1	Cleaner	1	25
1	Foreman	2	50
1	Manager	4	00
		39	05

	Crushing Plant Night Shift		
1	Dynamo Tender	1	75
1	Engineer	3	00
4	Chainers	5	00
2	Top men	2	75
1	Giant + Int	1	35
1	Roller feed	1	35
1	Bottom Elev 1	1	25
3	Small rolls	4	20
2	Crane	3	50
1	Dryer stack	1	50
1	" " asst	1	25
1	" Fireman	1	75
1	Conveyor 3	1	75
1	1+2 Conveyor	1	40
1	Cleaner	1	25
1	Foreman	2	50
		35	05

TOP: A worker stands atop the giant crushing rolls at the Ogden ore milling plant, 1895. BOTTOM: In this 1897 pocket notebook, Edison listed daily salaries for crushing plant employees. Managers earned $4 per day ($112 today), and cleaners were paid $1.25 ($35 today).

To increase the capacity of the new ore milling plant, Edison replaced the inefficient jaw-type crushers with a set of thirty-five-ton crushing rolls, each five feet long and six feet in diameter, which could crush rocks weighing as much as six tons. Edison calculated that it would be less expensive to crush rock with machinery powered by coal at $3 per ton than blasting the rock with dynamite at $250 per ton.

Edison automated as much of the plant as possible. Instead of employing workers to move iron ore by hand, the plant used steam shovels to scoop up the rocks, and traveling cranes loaded them onto rail cars, which brought them to the crushing rolls. A crane then dropped the ore between two giant rolls that turned in opposite directions. After the ore passed through the giant rolls in the crushing house, conveyor belts and elevators moved the ore through a series of successively smaller crushers that reduced the rock to fine particles. The ore was dried before conveyors moved it to the stock house.

From the stock house, conveyor belts transported the ore to the separating mill, where it passed through a series of screens and magnets to separate the iron from sand. Wasting

Edison installed a six-ton electric crane at the Ogden mill to move giant boulders in 1893. Behind the crane are the stone power house, the crushing mill, and part of the bucket conveyor shed.

nothing, Edison recovered the sand and sold it as a building material. Phosphorus on the concentrated ore was removed in a dusting chamber (the dust was also recovered and sold as a paint additive). After de-phosphorizing, the ore was molded into briquettes and loaded onto rail cars.

Construction of the giant crushing rolls, a critical component of the ore milling plant, was completed in early 1894. When Edison conducted his first test of the rolls in April, he discovered a number of flaws. The wood foundations supporting the rolls were not strong enough to prevent them from misaligning, which jammed the shaft bearings. The pulleys and belts driving the rolls broke. When a large rock was dumped into the rolls at an improper angle, it merely jumped up and down atop the crusher. Thus, Edison had to spend the remainder of 1894 and the first six months of 1895 redesigning the rolls.

The problems didn't end there, though. In late 1895, Edison's workers discovered that the links controlling the plant's elevators had deteriorated, requiring the redesign and reconstruction of the mill's entire elevator system. Further, lack of funds prevented Edison from

Edison patented a bucket conveyor system to move crushed ore around the plant. Preventing dust from clogging the rollers and bearings was a difficult technical challenge.

working on the ore milling plant during the first six months of 1896. Construction work resumed at Ogden only on a limited basis in the latter part of the year because the endeavor was still short of cash.

DESPITE THE HARDSHIPS and long hours—Edison typically clocked sixteen- to eighteen-hour days, except on Sunday, when he returned to West Orange—the inventor relished his time at Ogden. Former mine employee Dan Smith recalled that "these were happy days for Mr. Edison—Fine air, plenty of work, good food." The challenges of solving complex technical problems free from investor interference and other distractions were only part of the appeal. Edison and his workers (as many as 2,000 at one point) formed a community that worked, lived, and played together in what Smith called "a lovely setting of woods with gorgeous chestnuts and other trees."

Edison's workers and their families lived in comfortable bungalows equipped with electric lights. The men had little time to keep gardens, but some raised pigs and cows. To amuse themselves, workers staged boxing matches and rattlesnake and cockfights. The mill workers, Smith remembered, "liked the comfortable homes, the pleasant community life. Life might have seemed lonely to an outsider but it wasn't. Even the women were sorry to leave when the great experiment was over."

According to Smith, Edison's personal courage inspired loyalty. "He would never send a man anywhere that he would not go himself," Smith recalled. "Often the men would remonstrate with him for going into dangerous places." The ore milling plant was a hazardous work site. In August 1892, five workers were killed and twelve injured when a partially constructed stock house collapsed. Edison supervised the rescue of injured employees buried under heavy wood timbers.

Benjamin Odell, an electrical engineer who worked at Ogden, noted that the men liked Edison "because he was one of them. He ate with them, slept with them. He loved to swap lunches with them. He preferred the workmen's fare of hard-boiled eggs & corned beef to chicken."

Edison's love of practical jokes also endeared him to his workers. On one occasion Edison entertained a visiting titled Englishman, who was, as Smith described him, "a bit fussy about the dust from the mine on his clothes." When Edison became bored with the visitor, he told Smith, "Let's initiate him. When I get him under that platform you dump that chalk over him." Smith followed orders. "When the Britisher arrived at the fatal spot, over went the chalk. The Englishman's eyes, nose and mouth were filled & he had to be carried out half stifled." Edison simulated a "towering rage and swore like a pirate," while the workers pretended as if they didn't know anyone had been under the platform.

With Edison's assent, workers also mocked his reputation. When too many visitors showed up at the mill, an employee would ask Edison to come out of his office and "walk up & down and show 'em what genius looks like." Obligingly, Edison would "saunter up & down a few times & then turn his back on the lion hunters who would go away satisfied."

Edison outside his office at the Ogden mill, 1895.

At the end of a long day, after the evening meal, Edison gathered his men to talk about work problems in sessions he called "going to school." Edison, Smith remembered, "loved to rag his men, to haze them with questions—just to see if they really knew what they were doing." During these evening meetings, Edison sat in the background, puffing on a cigar, while the men discussed and debated technical issues. The inventor appeared inattentive until someone raised an interesting argument, at which point he would join the conversation. Edison, according to Smith, "always could hear what he wanted to, or when you didn't want him to." He expected his workers to argue with each other and with him. If he won an argument, he would pull out a pair of large donkey ears and prance around the room. If an employee scored a point, he would get to wear the ears. When the oral history interviewer asked Smith what happened if you did not go to school, Smith answered grimly, "We *went.*"

X-RAYS

Edison's laboratories did not pursue scientific research. As he told a Brooklyn newspaper in 1888, "I am not a scientific man, I am an inventor. . . . A scientific man busies himself with a theory. He is absolutely impractical. An inventor is essentially practical." Nevertheless, Edison kept informed of scientific advances, and when news of German scientist Wilhelm Röntgen's discovery of X-rays reached the United States in January 1896, he quickly began experiments to investigate the phenomenon. Optimistically, he told reporters that he could invent a practical X-ray apparatus in "two or three days."

While studying electrons emitted by a vacuum tube, Röntgen noticed that the electrons caused a nearby fluorescent screen to glow. Röntgen eventually produced an image of the bones of his wife's hand, but producing X-rays required a stable vacuum and precise voltages—a consistency that was difficult to achieve with available vacuum tubes.

Edison's X-ray experiments focused on designing a more reliable vacuum tube. Relying on the expertise of his chemists and glassblowers and on his own experience working with vacuum tubes during his electric light research, Edison designed a marketable X-ray tube. The laboratory staff also tested the effect of X-rays on different substances and made X-ray images on photographic plates. At the request of newspaper publisher William Randolph Hearst, the lab attempted, without success, to X-ray a human brain.

To reduce the time it took to expose an X-ray image on photographic film, Edison's experimenters began work on a "real-time" X-ray image viewer. This involved identifying the substance that produced the best fluorescence. Tests of more than 1,300 chemicals revealed calcium tungstate as the most fluorescent material.

Edison touted the medical benefits of his X-ray viewer, which he named the "fluoroscope." On March 26, 1896, a press release announcing the invention noted, "It will be invaluable to surgeons in locating the seat of injuries." For several years, the Edison Manufacturing Co. sold X-ray "outfits" that included vacuum tubes, fluoroscopes, and accessories to doctors and hospitals.

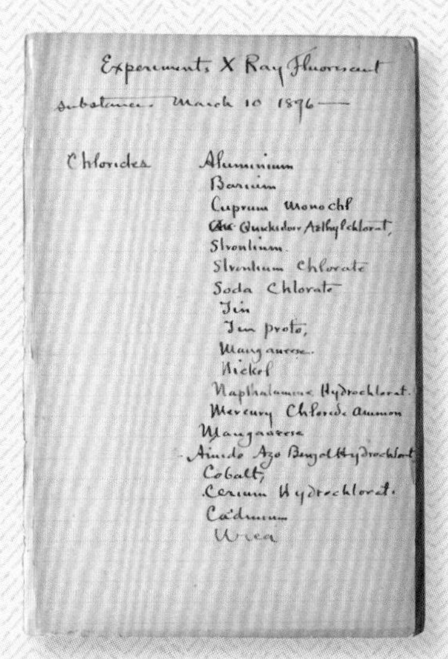

Experiments X Ray Fluorescent
Substances March 10 1896
Chlorides Aluminium
Barium
Cuprum Monochl
Strontium
Strontium Chlorate
Soda Chlorate
Tin
Tin proto,
Manganese
Nickel
Napthalamine Hydrochlorat.
Mercury Chloride Ammon
Cobalt,
Cerium Hydrochlorat.
Cadmium
Urea

TOP: Page from an Edison notebook, listing chemicals tested during the X-ray experiments. BOTTOM: German physicist Wilhelm Röntgen's experiments on vacuum tubes led to the discovery of X-rays.

By the fall of 1896, news reports that X-rays might play a role in curing blindness competed with accounts of the health risks of prolonged X-ray exposure. Weeks of experimenting with X-rays caused stomach and eye problems for Edison, and Clarence M. Dally, his principal researcher on the project, eventually succumbed to cancer in October 1904.

The tension between the benefits and risks of X-rays emerged in the medical treatment of President William McKinley, who was shot on September 7, 1901, while attending the Pan-American Exposition in Buffalo, New York. The president's secretary called West Orange and asked Edison to send his best X-ray machine and technicians so that doctors could locate a bullet lodged in McKinley's abdomen.

The X-ray machine and two Edison employees arrived at Buffalo on September 8. Doctors wanted to test the machine on another person before using it on the president, but they needed someone of McKinley's size. A local doctor with a fifty-six-inch waistline reluctantly agreed to the test. The test succeeded, but McKinley's doctors, concerned about side effects, decided not to expose him to X-rays. The bullet was never located, and McKinley died of a gangrene infection on September 14.

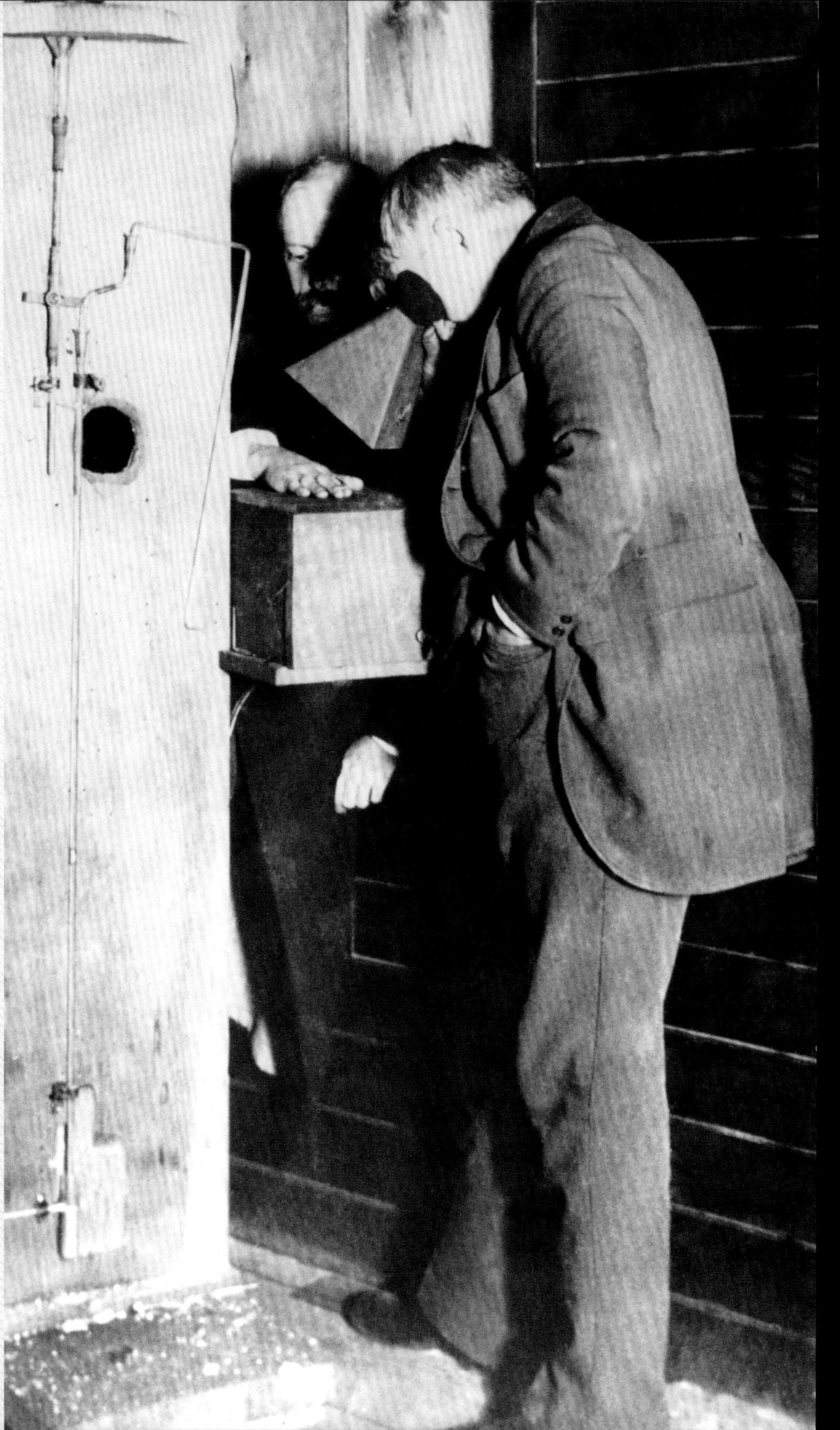

RIGHT: Edison peering into a fluoroscope in 1896.

The oral histories left by Dan Smith and Benjamin Odell portray Ogden as an idyllic work environment, but there were also hints of the strife that plagued labor-management relations in the late nineteenth and early twentieth centuries. In August 1896, machine shop workers, demanding time and a half for more than ten hours of work and double time on Sundays, threatened to strike. Edison fired all of the employees except for a few workers before the machinists could hold a strike vote.

IN EARLY 1897, the inability of the giant crushing rolls to efficiently break up large rocks was Edison's biggest challenge. He improved the crushing rolls by adding steel plates with four-inch-high knobs, and tests were promising, but some of the larger rocks stripped the bolts from the plates. After Edison found a way to secure the bolts in the summer of 1897, he told

Employees of the New Jersey & Pennsylvania Concentrating Works in 1894. Edison expected automation to reduce the number of workers he needed to operate the Ogden mill, but in the mid-1890s there were about 400 employees on the payroll.

Mallory, "That damn Giant Roll problem is at last solved, the rolls will now crush anything that will go into the hopper."

Nevertheless, during the late 1890s, Edison and his staff faced other engineering problems and equipment failures that prevented the plant from operating at full capacity. These technical issues required constant attention, prompting Edison to design and redesign machines for crushing, conveying, screening, separating, and drying iron ore.

The improvements were expensive. Of the $2.5 million invested in the ore milling experiment ($69.1 million today), Edison spent $2,174,000 of his own money. Work at Ogden proceeded irregularly during 1898 and 1899, as Edison and Mallory scrambled to find money to pay the plant's expenses. Edison drew funds from his two profitable companies, the Edison Manufacturing Co. and the National Phonograph Co. According to Mallory, "Mr. Edison and I would figure up the probable profits for the next few months and proceed to spend them and then later on go back to the manager of the Phonograph Co. and other company and tell him that we had these debts, which had to be paid." This angered the manager of the National Phonograph Co., William Gilmore, who needed company profits to expand the phonograph business. This continued until 1900, when Edison had expended all of the money available to him. Reluctantly, he closed the Ogden mill in September 1900.

The ore milling story did not end at Ogden, however. In response to international interest in Edison's ore milling process, a group of investors organized the Edison Ore-Milling Syndicate, Ltd., in June 1898. The syndicate, created to control Edison's foreign ore milling patents, initially planned to license the ore milling technology outside of the United States and Canada on a royalty basis. It soon decided, however, to operate its own ore milling plant on the west coast of Norway, 200 miles north of Trondheim. Edison believed that an ore milling plant at the site, known as the Dunderland tract, would "pay an interest on 50 million dollars if done on a large scale."

The Standard Construction Corp., Ltd., organized in February 1902, constructed the mill, rail lines, docks, and other plant facilities at the remote location. The Dunderland Iron Ore Co., Ltd., established in the spring of 1902, operated the plant.

The operators at Dunderland experienced the same problems that Edison had faced at Ogden, including dust-clogged machinery, but by 1908 the plant produced 800 tons of ore per day and had shipped 30,000 tons of briquettes. Declining iron ore prices, however, prevented the plant from operating at a profit. In 1909, it slipped into bankruptcy.

LLIPSBURG · DEVELOPMENT · COR
PHILLIPSBURG
N · J
INGERSOLL · MONOLITHIC · HOUSE

PORTLAND CEMENT

"CEMENT AND STEEL ARE TO BE THE BUILDING MATERIALS OF THE FUTURE."

ONE OF EDISON'S STRENGTHS as an innovator was his ability to apply technologies from one industry to try to solve the problems of another. His decision at the end of 1898 to manufacture Portland cement using his ore-handling and crushing machinery was an attempt to salvage the more than $2 million he had invested in the ore milling project.

Cement is a mixture of limestone and specific proportions of silica, alumina, and iron oxide that is burned in a kiln at a high temperature. The kiln's heat fuses the material, forming a "clinker" that is then ground into cement powder. Concrete forms when water is added to a mixture of cement and an aggregate—typically sand or crushed stone.

There are two types of cement: natural and Portland. In natural cement, the correct proportions of lime, silica, alumina, and oxide occur naturally in mined rock, without the intervention of cement producers. A common construction material in the nineteenth century, natural cement was used to build the Brooklyn Bridge, the U.S. Capitol, the Erie Canal, and the pedestal of the Statue of Liberty.

To produce Portland cement, the minerals are combined in the correct proportions by a manufacturer. Invented in 1824 by English bricklayer Joseph Aspdin, Portland cement was stronger and cured (or set up) faster. Because manufacturers, not nature, controlled the combination of minerals, the quality of Portland cement was more consistent. It was also more economical because builders could mix a lower proportion of cement with sand to form a mortar. Aspdin named the cement "Portland" because its finished surface reminded him of stone found on the British island of Portland.

European producers began manufacturing Portland cement in the 1830s. In the

PAGES 162–163: Tract of cement houses constructed in 1935 in Phillipsburg, New Jersey. RIGHT: David O. Saylor, president of the Coplay Cement Co., developed the first process for manufacturing Portland cement in the United States.

United States, Portland cement manufacturing began in 1871 at David O. Saylor's plant in Coplay, Pennsylvania. Domestic production of Portland cement increased dramatically in the late nineteenth century, climbing from 42,000 barrels in 1880 to more than one million in 1896. Two years later, there were twenty-four manufacturers producing three and a half million barrels of Portland cement. Nevertheless, in the late 1890s, the cement industry had trouble supplying demand, resulting in price increases that encouraged existing cement companies to expand their capacity and new companies to enter the business.

Vertical Portland cement kilns of the Coplay Cement Co., preserved today at the Saylor Park & Cement Industry Museum in Coplay, Pennsylvania.

Edison already understood the technology that was required to crush large quantities of rock, and his laboratory had the expertise to analyze the chemical composition of minerals—a significant skill for producing Portland cement—but he needed background on the existing supply and demand for the popular binding material. In December 1898, he began studying the cement industry, asking former Menlo Park assistant Francis Upton to collect detailed information about the physical properties, price, and markets for natural and Portland cement. Sometime during the spring of 1899, Upton gave Edison a notebook filled with information about the leading cement manufacturers.

Edison planned to manufacture Portland cement on a large scale, lowering production costs by automating as much of the cement mill as possible. On June 6, 1899, he organized the Edison Portland Cement Co., capitalized at $11 million, and began searching for a location to build the cement factory. Edison focused the search near Stewartsville and Phillipsburg, New Jersey. Two cement companies, the Vulcanite Portland Cement Co. and the Alpha Portland Cement Co., already had plants in the area, which had abundant deposits of limestone. In December 1899, Edison began purchasing farmland in New Village, a small town three miles east of Stewartsville.

Because Edison had little experience with roasting kilns, he studied the sixty-foot rotary kiln used by cement manufacturers, discovering that it could roast no more than 200 barrels of cement a day and was not fuel efficient. With characteristic ambition, he decided to invent a kiln that could roast 1,000 barrels of cement in twenty-four hours.

To understand how kilns operated, the inventor constructed a small-scale working model of a kiln in the spring of 1899. He tested rotation speeds, the insulation properties of different interior lining materials, and, with 150 barrels of clinker bought from the Lehigh Portland Cement Co., the flow of material inside the kiln. By the spring of 1900, Edison had designed a new kiln. It was 150 feet long, which allowed increased production of clinker, and featured a compressed air system that allowed more efficient and precise control of the kiln's fuel (pulverized coal) and oxygen supply.

Working with models helped lower innovation costs because experimenters could test ideas and resolve design problems before producing the blueprints that machinists used to make a full-scale prototype. Edison used this approach again in July 1900, when his laboratory staff constructed a model of the entire cement plant to determine the arrangement of machinery.

Edison's 150-foot-long kilns increased the production capacity of his Portland cement plant.

Construction of the cement plant began in the winter of 1900–1901. The plant's site covered forty-five acres and would be connected to a 600-acre quarry by a one-and-a-half-mile-long railroad. Visiting the site on January 9, 1901, Edison told reporters that the plant "would be one of the largest in the country and would be equipped with all the latest appliances and modern machinery."

Edison described the mill's operation in 1901. At the quarry, two ninety-seven-ton steam shovels loaded limestone and cement rock onto railcars, each carrying seven tons. The railroad brought the cars to the crushing house, where electric hoists dumped the rock into giant rolls, the first in a series of crushers that reduced the rock from four-and a-half-foot chunks to three-eighths-of-an-inch pieces. The crushing rolls were designed to process 300 tons of ore per hour.

From the crusher, conveyor belts moved the stone to a fifty-foot-high dryer, capable of heating 300 tons of material per hour. During the transfer from the dryer to a long stock house, the rock samples were automatically removed every thirty seconds and taken to the assay department, where testers analyzed the chemical composition of the material.

After drying, conveyors took the crushed limestone and cement rock to the stock house, where it was stored (separately) in one of seven cones, each with a 1,500-ton capacity. A large exhaust fan at one end of the stock house and a heater at the other finished the drying process. Next, conveyors took the limestone and cement rock to the weighing house, where they were loaded into separate bins. Using information from the assay department, workers mixed the limestone and cement rock together in the correct proportions to make Portland cement, transferring the material to a small stock house to await delivery to the grinding mill.

Rough sketches, like this one of the cement plant's crushing, grinding, and screen house, allowed Edison's staff to conceptualize the plant's operation before committing a design to a working model.

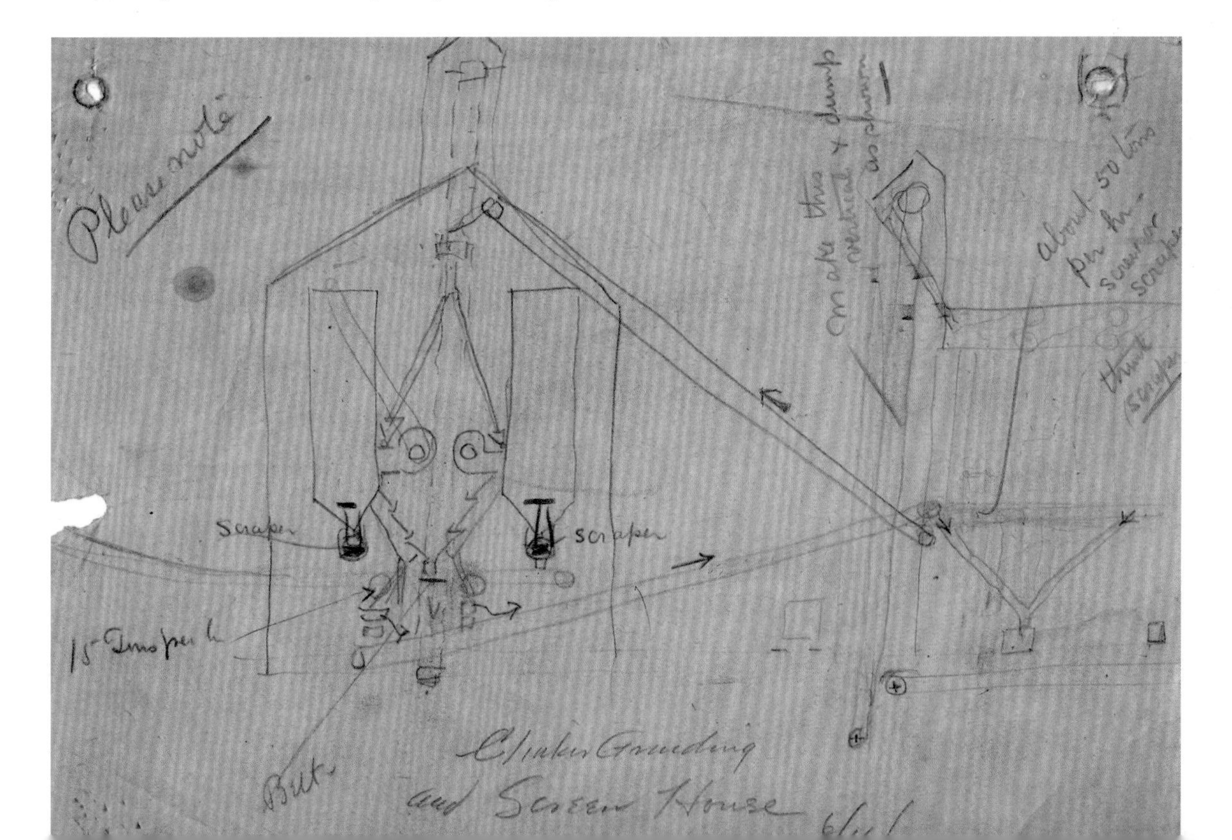

In the grinding mill, a series of crushing rolls ground the cement rock and limestone together into a fine powder, which would soon arrive at the rotary kiln to be roasted to form clinker. Conveyors took the clinker to another series of crushing rolls for yet more grinding. As the clinker passed through the rolls, screens captured large pieces that needed to be ground further. The finished cement was stored in another stock house, where two men and one boy used machines to bag and load the cement onto boxcars at the rate of 400 bags per hour.

WHEN WORKERS COMPLETED CONSTRUCTION of Edison's New Village plant in the summer of 1902, it was the largest cement factory in the world. Constructed at a cost of $1.5 million, it included twenty-seven buildings that covered an area a half mile long and a quarter mile wide.

The plant did not begin shipping cement until early 1903. Production was delayed because, as Edison's plant engineer recalled, "it was necessary to make a great many trials and experiments on the kilns and other machinery before we were able to manufacture cement commercially." Edison and his team resolved most of the plant's technical problems, but they experienced difficulties in getting the kiln to operate at full capacity.

Edison expected the kiln to roast 1,000 barrels of cement per day, but by the end of 1902 it produced only 500 barrels daily. Dissatisfied with these results, the inventor cajoled his plant engineers to increase kiln capacity. He also continued to test and modify equipment to reduce operating costs.

An explosion in the coal-grinding plant in March 1903 caused the death of several workers, including the plant's first engineer, Edward Darling. Edison closed the plant in order to redesign the coal-grinding machinery, taking the opportunity to improve other plant operations. Company stockholders, concerned about further expensive delays, urged Edison to use existing coal-grinding

Edison Portland cement was used in the construction of Yankee Stadium, the Traymore Hotel in Atlantic City, New Jersey, and the New York City subway system.

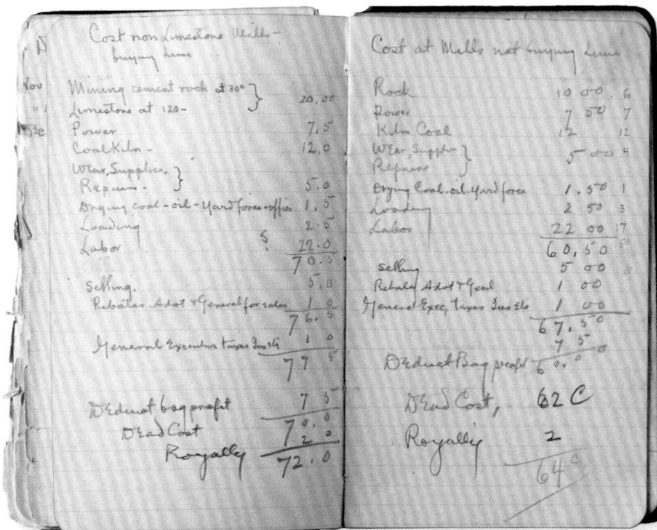

LEFT: This ad focused on Edison's contributions to the Portland cement industry and publicized his contributions to the New Jersey economy. RIGHT: In developing his inventions, Edison paid close attention to his costs. In this pocket notebook, he calculated the cost of limestone for the cement plant.

machines, but Edison believed strongly in the value of experimentation for improvement of the production process. As he told William Mason, the new plant engineer, "the only way to keep ahead of the procession is to experiment. If you don't, the other fellow will. When there's no experimenting there's no progress."

The Edison Portland Cement Co. did not become profitable until the 1920s. Demand for cement in the early 1900s had encouraged several new producers, such as Edison, to enter the business, but the industry soon developed an overcapacity problem. Cement sales declined in the economic downturn following the Panic of 1907, forcing Edison to close the plant for part of 1908. In 1913 Edison remarked, "Cement biz for five years has been a losing game." Edison closed the mill again from December 1914 to April 1916 because of declining cement prices.

ASIDE FROM SHUTTING DOWN OPERATIONS from time to time, Edison responded to low cement sales by looking for ways to encourage consumption. The Edison Portland Cement Co. issued color booklets instructing farmers how to make concrete watering troughs, silos, fence posts, and milk house floors. Other booklets promoted using Edison Portland cement for sidewalks, street curbs, septic tanks, and flowerpots. Edison cast decorative art pieces out of Portland cement to demonstrate its versatility. At the 1908 convention of the National Association of Cement Users, he exhibited an 18-inch by 14-inch by 4-inch bas-relief Indian head medallion, which is now displayed in the library at the West Orange laboratory. In 1911, Edison dabbled in making furniture out of cement and even tested cement phonograph cabinets. In a more conventional use, Edison's Portland cement was employed to construct a highway between Stewartsville and New Village, New Jersey.

In a June 1901 interview with the journal *Insurance Engineering*, Edison proposed constructing houses out of cement. "Cement and steel," he declared, "are to be the building materials of the future." He predicted that skyscrapers would be built with frameworks of steel and walls of cement, and that all housing for workers would be made out of "poured" cement. Building contractors could form floors and walls out of cement poured into wooden molds, erecting them quickly on the site. The cost, Edison believed, would be less than traditional wood construction and require only a few workers and a supervisor. There would be no need for carpenters or other skilled trades because everything in the house would be made out of cement, including the floors and stairs.

Edison experimented with making phonograph cabinets and decorative pieces out of Portland cement.

Edison also predicted that cement would become the preferred building material as its price dropped. "When the price gets to be $1 a barrel, or $5 a ton, and people know they can get it for that, there will be enormous quantities of it used." Cement houses could be built in a few days, and the rent for workers would be affordable—no more than seven or eight dollars per month ($191 to $218 per month today). When asked about the fire risks of cement houses, Edison replied that cement houses were safe and predicted that fire insurance companies would go out of business.

Edison was not the first to propose cement houses. In 1879, William Evans Ward used reinforced concrete to construct a large mansion in Port Chester, New York. Portland cement manufacturer David O. Saylor used cement to build a row of two-story houses for his workers in Allentown, Pennsylvania. In 1902, Harry Alexander Taylor—a stonemason from Jackson, Michigan—invented a molding system for building cement houses, and, in 1903, a builder from Grand Rapids, Michigan, built several single-family houses using reusable molds.

The idea of using new, less-expensive building materials to construct affordable houses for working people appealed to turn-of-the-century social reformers who were concerned about the "tenement problem." Increased immigration from eastern and southern Europe in the late nineteenth century had led to overcrowded and unsanitary housing conditions in the nation's large cities. Seeking affordable housing, many immigrants were living in substandard tenement housing. According to a 1901 New York State report on tenements, in 1864, 486,000 people lived in 15,511 tenements. By 1900, 1,585,000 people lived in 42,700 tenements. As the report concluded, "Of all the great social problems of modern times incident to the growth of cities, none is claiming public attention in a greater degree than that of the housing of the working people."

To increase sales, Edison promoted new uses for Portland cement. These ads encouraged farmers and do-it-yourself homeowners to consider Portland cement as a building material.

One solution to the tenement problem was developing affordable housing for working families in suburbs, but this was not a feasible option for most tenement residents. "A family which now pays from $12 to $18 a month for its apartment in a tenement house must be able to pay at least $20 a month for a separate house in the suburbs," the New York State Tenement Report noted.

Edison did not pursue the cement house idea until 1906, when he announced his plan to construct cement houses in one continuous pour, using reusable iron molds. His first effort to cast a cement structure may have been a chicken coop built on the Glenmont grounds. No evidence of this house has been found, except for a newspaper article in which Edison was quoted: "Members of my family laughed at me when I told them I was going to make a chicken coop out of concrete but they are not laughing at me now."

In the winter of 1907, Edison hired New York architects Horace B. Mann and Perry R. MacNeille to design a cement house. Mann and MacNeille submitted plans for a two-story, two-family house. The exterior included arched openings, pilasters, cornices, and other elaborate Queen Anne–style architectural elements.

During the summer of 1907, Edison's staff began building a quarter-scale model of the design on the third floor of the West Orange laboratory. Edison showed the model and introduced his plan for mass-producing cement houses to members of the American Electrochemical Society on October 18. The next day, newspapers published descriptions of Edison's "system for building a house complete in 12 hours." Edison told the *New York Times*, "I am going to fashion cast iron molds for the entire house. This outfit will

Edison with a model of his redesigned cement house, ca. 1910.

cost $30,000 for a house of this design. All the builders will have to do is to put it up and pour in the concrete. Then they will allow six days for settling and drying and the family may move in." Touting the durability of the cement house, Edison told the *Times*, "The man who owns the house can let his children hack at it with hatchets and axes."

"IF I SUCCEED . . . THE CEMENT HOUSE WILL BE MY GREATEST INVENTION. IT WILL TAKE FROM THE CITY SLUMS EVERYBODY WHO IS WORTH TAKING."

Henry Phipps, Jr., a wealthy philanthropist, brought a group of architects and builders to the West Orange laboratory in November. Phipps was interested in the tenement problem and, in 1905, had donated $1 million to develop affordable worker housing in New York City. Phipps did not invest in Edison's cement house, but he offered comments that raised interesting questions about the design of affordable housing. Concerned that excessive decoration would raise the cost of Edison's cement house, Phipps encouraged a simpler, less ornate design. Edison disagreed:

> I think it would be an error not to build the most beautiful house that is possible. . . . If the slums of cities can be depopulated only by building rows of plain box houses then that is the way to do it. But if the same can be done by making these houses the most beautiful that art can create, then I think it would be a sad mistake not to make them this way.

Edison believed that the cement house would have to be attractive *and* affordable.

In March 1908, civil engineer Edward Larned and Association of American Portland Cement Manufacturers secretary Percy H. Wilson published an article in *Cement Age* criticizing Edison's cement house system. Like Phipps, they concluded that the design was too complicated for economically efficient construction, doubted that Edison could make cement flow and settle properly in the iron molds, and claimed that Edison had considerably underestimated the cost of the molds.

Edison may not have appreciated the outside criticism, but his own staff was not happy with the Mann and MacNeille design, noting such problems as the lack of windows in

ABOVE: Blank spaces on advertising placards allowed local cement dealers to add their names and addresses. BELOW, RIGHT: Promotional calendar for Edison Portland cement, 1920.

the attic, no door in the rear of the house, and not enough windows at the front of the house. As a result, Edison set aside Mann and MacNeille's plan and hired two engineers, George Small and Henry Harms, to design a new, simpler house, as well as the cast-iron molding system.

During the summer and fall of 1908, Small and Harms redesigned the Edison cement house, replacing the two-family dwelling with a two-story, single-family house with six rooms, one bathroom, a two-room attic, front and back porches, and a basement. They also experimented on a critical problem related to the flow of cement inside the iron molds. When poured, the cement had to flow evenly through the entire mold system and remain suspended until it set and dried. Their solution was to mix clay and other additives with water before it was mixed with the cement and sand. This created a colloidal suspension that kept the cement aggregate dispersed during the drying process.

Edison did not intend to build cement houses himself. Instead, he offered contractors royalty-free use of his molding system. In a circular drafted to respond to numerous inquiries from the public, Edison wrote,

> I have not gone into this with the idea of making money from it, and will be glad to license reputable parties to make moulds and erect houses, without any payments on account of patents. The only restriction being; that the designs of the house be satisfactory to me, and that they shall use good material.

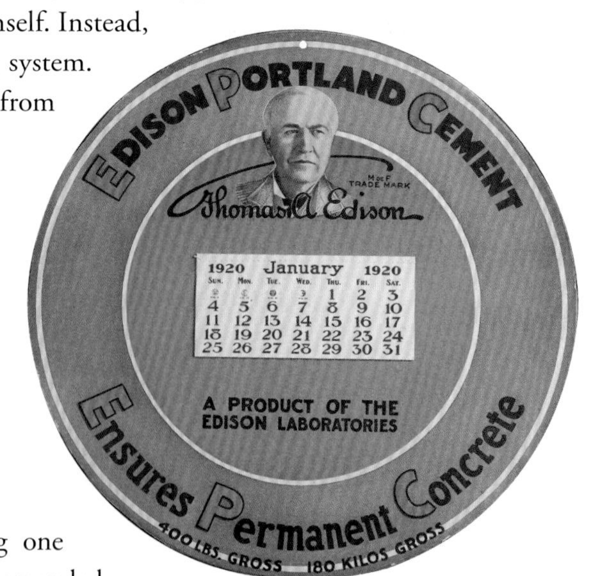

Because a set of molds cost $25,000, building one cement house from it was not practical. Edison recommended that builders work with six sets of molds to keep construction workers busy. Additionally, six sets would allow them to build ten to twelve houses a month. This limited the use of Edison's system to contractors who could afford the high set up costs.

Several developers used Edison's concepts to construct cement houses. In 1908 the American Concrete Co. constructed a cement cottage in Long Branch, New Jersey, and Frank Lambie, a retired shoe manufacturer, built an experimental cement house in Montclair, New Jersey. Lambie developed his own molding system, based on Edison's ideas, which he eventually used to construct 100 houses for the American Steel and Wire Corporation in Donora, Pennsylvania, during the First World War. In 1918 Charles H. Ingersoll, the manufacturer of the Ingersoll "dollar watch," a cheap, mass-produced pocket watch, constructed thirteen cement houses in Union, New Jersey, and at the end of the year began building cement houses in a development project in Phillipsburg, New Jersey. Ironically, although Edison's cement plant was only a few miles from the site, he did not supply the cement. Alpha Portland Cement Co. official Frank Coogan was also the president of the company managing the Phillipsburg project, and he made sure builders used Alpha cement.

TOP: Edison cement house molds and machinery under test at the West Orange lab in 1911. BOTTOM: Cement houses under construction in Union, New Jersey, 1919.

24
35010

10

STORAGE BATTERY

"I DON'T THINK NATURE WOULD BE SO UNKIND AS TO WITHHOLD THE SECRET OF A GOOD STORAGE BATTERY IF A REAL EARNEST HUNT FOR IT IS MADE."

AT THE TURN OF the twentieth century, the question of whether automobiles would be powered by gasoline, electricity, or steam remained open. Nearly 30 percent of the automobiles produced in the United States in 1900 were electric, and although Edison had predicted that automobiles would replace horses in 1895, he doubted that vehicles would be powered by electricity. Four years later, he changed his mind and began experimenting on storage batteries in the summer of 1899, spending ten years and nearly $2.5 million developing a marketable storage battery for electric vehicles.

There are two types of batteries: storage and primary. Both have positive and negative electrodes immersed in electrolytes—or solutions of acids, alkaline, or salts—and both produce electricity through a chemical reaction between the electrolytes and positive electrodes. The chemical reaction in primary batteries depletes the electrodes and electrolytes, which must be replaced. The reaction in storage batteries is reversible, and the cells can be recharged.

Primary batteries were used in the nineteenth century to power telegraph and telephone systems. They also provided power for electrical experiments in research laboratories. French physicist Gaston Planté invented the first lead-acid storage battery in 1859. In the early 1880s, Edison considered using storage batteries to store electricity during electric light central station off-peak hours, but ultimately decided that they were not practical. When the *Boston Herald* asked him in 1883 if he saw any hope for the storage battery doing legitimate work, he answered, "None whatever."

During the 1890s, the design of storage batteries had advanced far enough for electric utilities to start using them for power storage. This application may have convinced Edison that developing a practical storage battery was feasible. The first storage-battery-powered automobile was invented in 1894. However, because the batteries were heavy, difficult to maintain, and prone to corrosion and deterioration, they were not an ideal power source for vehicles. Edison began his storage battery research by outlining its technical requirements. To compete with lead-acid batteries, he would invent an easier-to-maintain, inexpensive, corrosion-resistant, lighter battery that would hold its electrical capacity for a long time.

Initial research focused on identifying materials to replace the electrolytes and electrodes in lead-acid batteries. Edison first considered using sodium hydroxide for the electrolyte, zinc for the positive electrode, and copper oxide for the negative. He soon settled on potassium hydroxide for the electrolyte but continued experimenting on electrode materials. For the negative electrode, the laboratory tested silver oxide, various copper compounds, and cobalt oxide before selecting nickel oxide. Cadmium was an early candidate for the positive electrode, but Edison concluded that it was too expensive and replaced it with iron. To increase the surface area between the electrodes, Edison ground the nickel oxide and iron into fine powders. This required the design of thin, flat pockets of nickel-plated steel, which held the iron and nickel inside the cell. Edison added graphite flakes to the nickel oxide and iron to improve conductivity.

By 1901, after conducting thousands of experiments on different combinations of metals and chemicals, the laboratory had settled on the basic components of the Edison storage battery: iron, nickel oxide, and potassium hydroxide. Identified as the Type E, each battery cell was encased in a steel container twelve inches high, six inches long, and four inches wide.

EDISON APPLIED FOR his first storage battery patents on October 15, 1900. Although the initial storage battery campaign ended by 1909, the West Orange lab continued improving the technology, and Edison submitted additional patent applications for battery modifications and manufacturing techniques. By the time he signed his last battery patent application on October 11, 1927, he had received 140 storage battery patents.

In May 1901, Edison organized the Edison Storage Battery Co. and established a battery factory in Glen Ridge, New Jersey. In June he sold his U.S. patents to the company for $1 million ($1,000 in cash, $999,000 in stock)—money he used to fund experiments and equip the factory. His marketing plan was simple: He would sell the batteries at discounted prices directly to automobile manufacturers, who were expected to design carriages to accommodate them. Consumers could purchase Edison batteries from the factory, but they were charged the list price: $10, $15, and $25 ($264, $396, and $659, respectively today), depending on the size of the cell.

Edison searched for his own sources of nickel to control raw material costs and to ensure reliable supplies. In

PAGES 176–177: Delivery trucks of the Chamette Laundry Co., equipped with Edison storage batteries, New Orleans, Louisiana, ca. 1920. RIGHT: Edison steel alkaline storage battery cell. "Built like a watch, but as rugged as a battleship."

September 1901, he sent his brother-in-law John V. Miller and a team of prospectors to locate nickel mines in Ontario's Sudbury mining district. In July 1902, Edison organized the Mining Exploration Co. of New Jersey, with the financial support of United States Steel Corporation executives, to widen the nickel search to Connecticut and other locations.

Edison in the West Orange library with his storage battery. Originally designed to power electric vehicles, Edison's storage battery eventually found wide use in railroad signaling, mining, marine transportation, and other light industrial applications.

Edison began testing the Type E battery in the spring of 1903. In April he wrote to Levi C. Weir, president of the Adams Express Co., "I shall test five automobiles between the factory and Morristown [N.J.] and when the batteries have run 5000 miles each at a high speed and do not show any deterioration and I am entirely satisfied that they are all right, I will then start the factories." In one of these tests, a light electric runabout equipped with twenty-one cells weighing a total of 322 pounds ran a sixty-two mile course between West Orange and Paterson, New Jersey, over hilly roads, reaching grades of twelve percent. Another test run covered eighty-five miles over muddy, rain-soaked New Jersey roads.

Edison subjected the storage battery to punishing road tests because, as he explained to New York banker William D. Sloane in August 1903, "The large carriage builders will go into the electric vehicle business on a considerable scale as soon as they are satisfied the battery is all right." Road tests not only gave Edison information to improve battery design, but also demonstrated the battery's performance to automobile manufacturers.

Edison was optimistic that his storage battery would, as the *Chicago Record-Herald* reported, "revolutionize the automobile business. . . . Mr. Edison himself expects to see his battery take the place of horses on delivery wagons of all kinds in cities, and he believes that in time the principle will be extended to the propulsion of street cars, railroad trains and steamships." The new battery, Edison wrote to Levi Weir, "will I think solve the problem of vehicle traction. It will have half the weight of those now in use, requires no attention, is fool proof, and has the merit of having none or very slight depreciation."

By 1904 several auto manufacturers, including the Pope Motor Car Co., Studebaker Brothers Manufacturing Co., the Baker Motor Vehicle Co., and the Electric Vehicle Co. had designed cars for the Edison storage battery. Edison recommended the Studebaker to William Sloane but preferred the Baker electric for women. "The Baker is a very light runabout, weighs 850 to 900 lbs and will go as far as a Studebaker but its generally used by ladies and young girls and its considerably cheaper. My daughter has used one constantly for 2 ½ years."

The Glen Ridge factory began manufacturing storage batteries in January 1903. Weekly production soon reached 200 cells, and initial battery sales were promising. One Edison Co.

A Bailey electric car in the West Orange lab courtyard in 1910, after a 1,000-mile endurance run. To test their durability, Edison subjected his storage batteries to punishing trials.

official remarked in January 1904, "The storage battery has been on the market since July last. We have had so many orders come to us, it has not been necessary to advertise it." In July 1904, Edison wrote, "We are getting along all right, we are manufacturing and selling all we can make, our business is now at rate of $300,000 per year."

In November 1904, Edison ordered the factory to stop battery production after he began receiving reports of leaking battery cans and loss of electrical capacity after several charges and discharges. He told Levi Weir, "At present I do not want to sell any more batteries than I am compelled to until I solve the leaking can problem."

The leaks were caused by chemicals corroding the solder used to seal battery cans. Edison solved the problem by replacing solder with a new welding process to seal the cans. The loss of capacity was a more difficult challenge. Edison noted in October 1904 that "after a long run some cells maintain their original capacity while others lose 30% although all made under the same conditions."

It took Edison and his experimenters several years to identify and resolve this issue. Ultimately, they discovered that the nickel oxide electrodes expanded during the charging process and contracted during discharge, which increased the cell's internal resistance and lowered capacity. Edison corrected this by designing a round electrode pocket to replace the Type E flat pocket. He improved the conductivity of the pockets by adding layers of nickel flake between the nickel oxide and by designing machinery to compact the layers of nickel under high pressure.

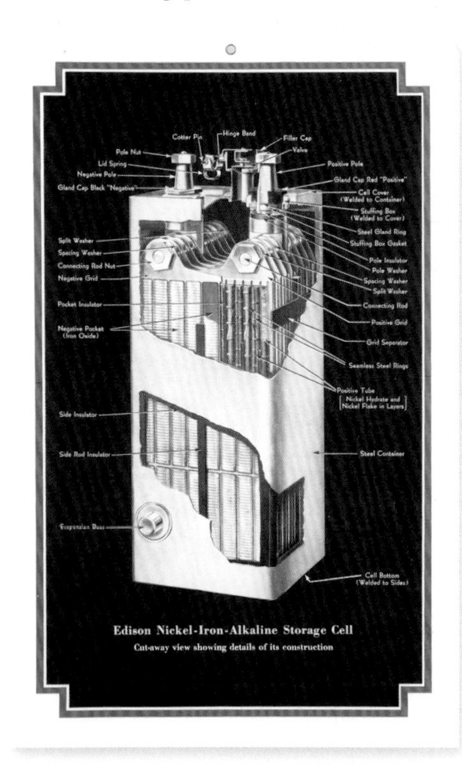

Cutaway of Edison storage battery cell. Each cell contains two sets of positive and negative plates. The plates support grids of tubes and pockets holding the active materials of nickel hydrate (positive) and iron oxide (negative).

EDISON BEGAN MANUFACTURING the redesigned storage battery, which he named the Type A, in July 1909. In correcting the Type E's defects, he had improved the battery's electrical capacity and increased the time a cell could hold its charge.

Edison expected a heavy demand for the new battery and began constructing a larger storage battery factory across the street from the West Orange laboratory to replace the Glen Ridge plant.

By 1910, however, the market for electric pleasure vehicles had changed. In 1908 Henry Ford introduced his low-priced Model T, which made gasoline-powered cars more affordable. Before the invention of electric starters, gasoline engines had to be hand-cranked—an especially difficult task on cold mornings. Consumers appreciated the convenience of starting electric cars, but gasoline-powered cars were faster, were less expensive to operate, and had a wider touring range.

When Edison checked automobile registration records in Trenton, he learned that 98 percent of the electric cars registered in New Jersey from 1899 to 1906 had been abandoned because of the unreliability and expense of lead-acid storage batteries. Electric vehicle owners complained about the length of time it took to recharge batteries and about the low mileage they could travel on a single charge. As William M. Lawrence, the pastor of the North Orange Baptist

TOP: Edison storage battery assembling department, West Orange, January 1915. BOTTOM: Workers operate tube-loading machines in the West Orange storage battery manufacturing plant. Along with new products, the West Orange lab designed machines to mass-produce them.

Church, explained to Edison in June 1910, "If I were asked in a word what is the main difficulty, I would say the uncertainty regarding the length of time the car will go on one charge. Sometimes this year I have not gotten more than eleven miles." Thomas W. Harvey, a doctor in Orange, New Jersey, complained about his electric car experience:

> I put in new batteries about every 2,500 miles. I ran the car about 2,000 miles a year, it cost me about $500 a year, but I had the satisfaction of knowing that only the wealthy can really enjoy life, and I took great comfort from the fact that even if a five thousand dollar limousine did whisk past me at a forty mile gait, when I was laboriously doing four miles an hour, still my ride was costing me so much more than the other fellow's that I could look down with disdain upon him.

Edison believed his new Type A storage battery would overcome these objections:

> The mileage is such with the new battery that man & wife & child can go out all day on one charge, on good roads here 150 miles is being done. They will probably displace the cheap gasoline car as people are finding that the repairs & upkeep of the later are really prohibitive & they seldom last four years. Whereas any lady can run the electric & the up keep as compared to gasoline is a mere nothing.

Automobile owners who preferred electric vehicles, however, tended to purchase the Exide, the lead-acid storage battery manufactured by Edison's principal competitor, the Electric Storage Battery Co. Although Edison's battery was more durable and less expensive to operate over the long run, it was more expensive to purchase than the Exide. Washington, D.C., lawyer James K. Jones explained that he could buy a 1911 Columbia with an Exide battery, "which will make about 75 miles on a charge for $1,750, but the same car with the Edison battery will cost about $2,400. It would appear to make the Edison battery cost $650 more than the Exide battery, though you get more mileage out of the Edison battery and it would probably last longer." Although Edison maintained that his batteries were less expensive to operate and service, they contained complicated components that were more expensive to manufacture. As a result, Edison could not compete with lead-acid batteries on price. Moreover, the Exide battery manufacturers guaranteed the quality of their batteries. Edison offered guarantees only to commercial truck owners because the company had no control over the installation of batteries on personal automobiles.

Edison aggressively searched for alternative storage battery markets. Pleasure automobile owners were more sensitive to battery purchase costs than commercial vehicle owners, who could amortize those costs over time. In August 1910, he asked P. V. DeGraw, the fourth assistant postmaster general, if the U.S. Post Office would be interested in an electric delivery truck. "If a light high speed (15 mph) delivery wagon carrying from 600 to 1,000 lbs could be made not costing more than 850 dollars & of extreme simplicity, employing electric motor & storage battery that would last for years, could the Post Office dept make use of such a vehicle in cities."

The electric delivery truck was not a new idea in 1910. According to *The Grid*, an Edison Storage Battery Co. trade publication, Edison had recognized the potential for electric delivery wagons back in the spring of 1900, while waiting for a ferry to take him back to New Jersey after a business trip to Manhattan. Edison experienced difficulty reaching the Cortlandt Street ferry because of congested traffic. "For two hours he stood watching this conglomeration of loaded trucks, irate teamsters and fretting horses." As *The Grid* related, Edison used the time to jot some ideas in his notebook: "Problem—narrow streets. Comparatively large street area

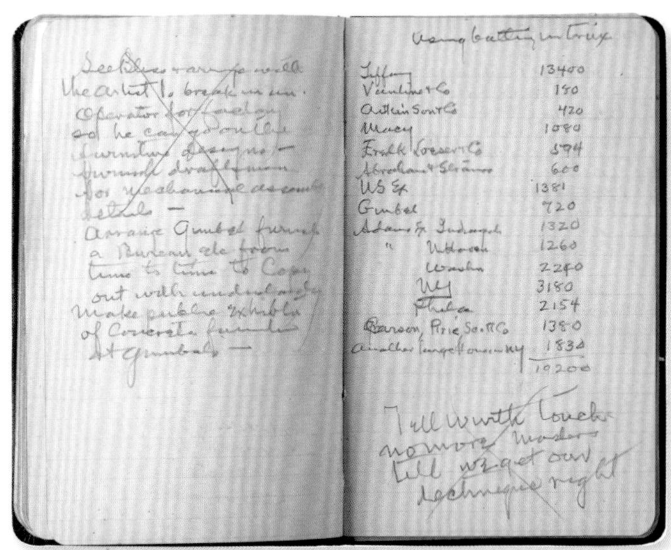

TOP: The Adams Express Co. tested Edison storage batteries on twenty-two trucks in Indianapolis for five years. Each year, the trucks ran an average of 40,000 miles at a cost of two and a half cents per mile. BOTTOM: Edison's list of the companies using storage batteries in their delivery trucks, including Tiffany, Macy's, Abraham & Straus, and Gimbels. On the opposite page he jotted notes about exhibiting cement furniture at Gimbel's department store in New York City.

covered by horse drawn vehicles. Slow speed. Limited loads. Congestion. Resulting delay and expense therefrom." Under these notes, he wrote, "Solution—electrically driven trucks, covering one-half the street area, having twice the speed, with two or three times the carrying capacity."

By March 1905, 250 commercial trucks used Edison's storage battery. Tiffany & Co. in New York operated twenty-one delivery wagons powered by Edison batteries in the fall of 1906. In crowded cities, electric delivery trucks had a few advantages over gasoline trucks. Electric wagons could be operated all day and garaged overnight for recharging. Because they were easier to start, electric delivery wagons were suitable for frequent delivery stops. When the first electric starters were introduced in 1912, electric trucks lost this advantage, but the Edison Storage Battery Co. continued to market batteries for delivery vehicles into the 1920s.

Edison also worked with Ralph Beach, a former General Electric engineer, to develop storage-battery-operated streetcars, and Beach organized the Federal Storage Battery Car Co. in 1910. In March of that year, a Beach streetcar equipped with Edison batteries began operating on a line in Manhattan. On September 25, 1912, three battery-powered Beach railcars, carrying 140 passengers, traveled from Pennsylvania Station in New York to Long Beach, Long Island. The train completed the 25.6-mile trip in fifty-six minutes.

Beach cooperated with Cornelius J. Field, who designed an electric omnibus capable of carrying twenty-eight passengers. "It is not a beauty, it is a very husky car & very practical," Field told Edison. "It made a run from New York to Atlantic City, 135 miles & back 150 miles without any trouble, speed about 10 miles an hour." Edison opted not to work with Field because of unfavorable credit reports.

One of the ideas that Edison entertained involved using the motion of ocean waves to charge batteries on lighted navigation buoys. He also liked an idea suggested by the owner of a Norwegian waterfall: use hydropower

Indian Scout motorcycles equipped with Edison storage batteries. The Hendee Manufacturing Co. (renamed Indian Motorcycle Co. in 1928) manufactured the Scout from 1920 to 1946.

to charge storage batteries on ships and then sell the electricity in England. Edison's chief engineer, Miller Reese Hutchison, asked the Electra Cycle Co. in Detroit if they were interested in using Edison's battery in their electric motorcycle. Electra put Hutchison off, claiming they were undergoing reorganization, but the Hendee Manufacturing Co. of Springfield, Massachusetts—the makers of Indian motorcycles—was interested in Edison's battery. On November 17, 1911, company president George M. Hendee asked Edison for a meeting to "talk over the question of Edison battery and electric motor for motive power on motorcycles." Edison agreed, although no record of an actual meeting exists.

Edison examines a storage-battery-powered searchlight in the West Orange lab courtyard, June 10, 1915.

Edison had more success marketing his storage battery as a component of his Country House Lighting system. Designed to provide electricity for homes and farms too remote to buy power from utilities, this system included a gasoline engine, a generator, a switchboard, and storage batteries. The board of directors of Mount Vernon, George Washington's Virginia home, purchased an Edison lighting system in 1920. The board, "after considering the dangers of oil lamps used up to the present, decided to install electric lights for the protection and preservation of this wonderful old home." During installation the local contractor concealed the electric wires. For areas inaccessible to wires, Edison provided Mount Vernon with several six-volt portable light batteries to power table lamps.

Edison also attempted to promote the storage battery for submarines. The submarine was an effective weapon in the arsenals of Great Britain, France, Germany, and Russia, but 200 sailors were killed in explosions, accidents, and collisions between 1903 and 1915. Edison believed that many of these accidents had been caused by leaking sulfuric acid from lead-acid storage batteries. When mixed with seawater, sulfuric acid forms lethal chlorine gas. His

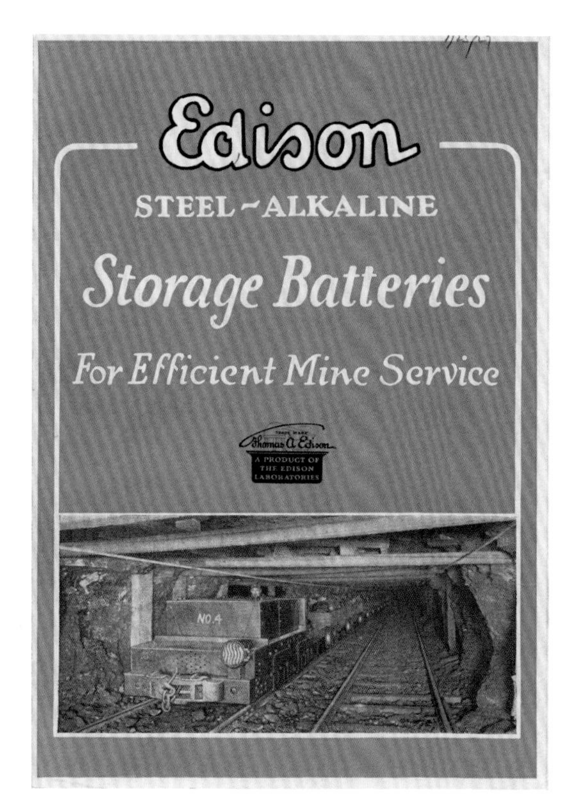

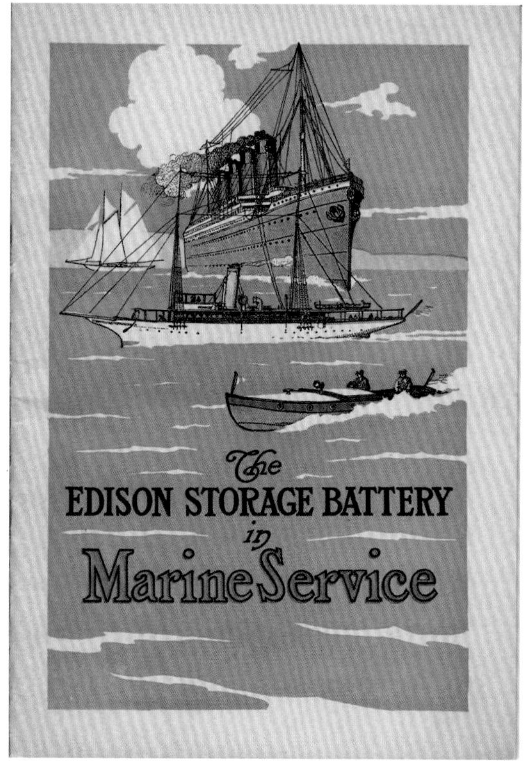

Edison promoted his storage battery as a safe, reliable source of electricity for mines, boats, and ships. According to a 1915 Edison Storage Battery Co. Bulletin, the presidential yacht *Mayflower* installed 100 Edison cells for emergency radio and lighting.

storage battery, which used no acid, would be a safer alternative. Hutchison told Edison, "Those submarine boys are dead crazy to get Edison batteries on the boats . . . you cannot well blame them when they are taking their lives in their hands every time they make a dive owing to the constant threat of chlorine gas."

The U.S. Navy agreed to test Edison's storage battery and installed them on the E-2 submarine at the Brooklyn Navy Yard in November 1915. However, tests revealed that the Edison cells released explosive hydrogen gas during the recharging process. On January 15, 1916, while the E-2 was in dry dock for the installation of a ventilation system, the submarine exploded, killing five men and injuring ten.

The explosion was a personal embarrassment for Edison who had claimed that his storage battery was safe. The navy's final report on the accident, sent to Congress in December

1916, did not specifically fault Edison, but concluded that no Edison storage batteries would be used in submarines until they were proven safe. Meanwhile, the navy refitted the E-2 with lead-acid batteries.

"A GREAT INVENTION WHICH FACILITATES COMMERCE ENRICHES A COUNTRY JUST AS MUCH AS THE DISCOVERY OF VAST HORDES OF GOLD."

DESPITE THE E-2 EXPLOSION and Edison's inability to market storage batteries for electric vehicles (his original goal), the Edison Storage Battery Co. successfully promoted the battery for a variety of industrial applications, including the powering of truck and automobile lighting and ignition systems, telephone switchboards, and emergency lights for factories, stores, and theaters. Edison's battery also provided electricity for ships and boats, radio sets, train lighting, and railway signaling and switching equipment.

Safety, durability, and reliability were strong selling points for a small battery Edison developed for miner safety lamps. These lamps illuminated dark mines while reducing the danger of gas explosions caused by candles, kerosene lamps, or other lights with open flames. Edison even produced a safety lamp for the mules used in mining operations. The locomotives used in mines to transport coal and other minerals were also operated with Edison batteries.

Additionally, the Edison storage battery was a lightweight, economical option for powering forklifts and tractors in factories, lumberyards, railroad terminals, and other industrial sites. As the Edison Storage Battery Co. boasted in January 1920, "Large industries, freight and steamship terminals and warehouses are finding that these storage battery trucks and tractors are effecting savings as high as 50% in handling costs besides speeding up organization and eliminating much unnecessary hand labor."

The Edison Co. continued to manufacture storage batteries in West Orange until 1960, when it sold the business to Edison's former rival, the Electric Storage Battery Co.

S.P. 192

11

EDISON IN WORLD WAR I

"MODERN WARFARE IS MORE A MATTER OF MACHINES THAN MEN."

EDISON WAS AN OUTSPOKEN ADVOCATE of military and industrial preparedness during the First World War. Recognizing that new technologies—including the submarine, machine gun, and airplane—were rapidly changing warfare, Edison told the *New York Times* in October 1915, "The soldier of the future will not be a sabre-bearing, blood thirsty savage. He will be a machinist." Future military conflicts, he believed, would be wars "in which machines, not soldiers, fight." In another interview with the *Times* several weeks later, Edison remarked, "Science is going to make war a terrible thing—too terrible to contemplate. Pretty soon we can be mowing men down by the thousands or even millions almost by pressing a button."

If recent inventions like the submarine and airplane raised security challenges for the United States, Edison believed that technology combined with industrial organization offered

PAGES 190–191: Edison with the officers and crew of the USS *Sachem*, 1917. The *Sachem* was a 217-ton yacht the U.S. Navy commissioned for Edison's research in August 1917. BELOW: Edison (passenger seat), Navy Secretary Josephus Daniels (rear seat, left), Miller Reese Hutchison (rear seat, right), and Charles Edison (obscured) arriving at the Brooklyn Navy Yard on October 12, 1914. Theodore Edison is at the wheel.

solutions. In May 1915, he outlined a preparedness plan based on the idea that military training and equipment procurement should be organized along industrial lines. He called for the stockpiling of airplanes, battleships, and munitions and for the recruitment of a large army of reservists trained by private industry. To develop new inventions quickly, he proposed the creation of military research laboratories.

As president of the Naval Consulting Board and a researcher for the U.S. Navy during the war, Edison had an opportunity to address these issues, but he faced a more immediate challenge when the war began in August 1914: a shortage of German and British phenol, or carbolic acid. Edison used carbolic acid to manufacture phonograph records, but because the chemical was also used to make explosives, the British and German governments embargoed its export.

Each day Edison's phonograph factory used one and a half tons of carbolic acid, a by-product of coal tar. As he told the editor of *Iron Age* in September 1914, "Carbolic acid is not obtainable in this country, our tar having scarcely any, so we are dependent on England & Germany. I am the largest user of carbolic here & the embargo put on shipments by England and the impossibility of getting any from Germany put me in a pretty tight space."

At first, Edison worked through political and diplomatic channels to reestablish his supply line. At his request, New Jersey senator William Hughes and Congressman Edward W. Townsend petitioned the State Department to intervene with the British on his behalf. Secretary of State William Jennings Bryan asked the British to lift the embargo and

THE fighting slogan in France, gathering inspiration and significance as the conflict grows more violent and more desperate, is "Carry On." On land, on sea, in the air, it rings sharp and clear.

Into the front line trenches comes the signal to charge. The company commander swings "over the top." At his heels, pushing and stumbling through the hell of "No Man's Land," come the boys. They gain a yard, five, ten, and the machine guns speak. The commander falls, but over his shoulder, above the din of battle, he shouts, "Carry On, Lieutenant!" So on and on, till every officer falls, and the grizzled old Sergeant sets his teeth and takes what's left of them on to victory.

"*Carry On*" must be our slogan here at home. We must "Carry On" to the utmost limit of our ability, to the last dollar of our resources, till *Victory* is won. Let us stand shoulder to shoulder—*buy* all the Liberty Bonds we can. Let us *keep* our Bonds and *save* to buy more.

Buy Liberty Bonds! Build More Batteries!
"Carry On!" "Carry On!"

BULLETIN NO. 19. EDISON STORAGE BATTERY CO., ORANGE, N. J.

Industrial trucks equipped with Edison Batteries in a big munitions plant in England. Note the women drivers.

Rushing Munitions

The batteries you make are used to make the wheels go in trucks in the large munitions factories of America and England and France.

More munitions are wanted all the time. So are more trucks to handle them.

It is up to us to see that there are plenty of batteries to equip the trucks.

YOUR PART IN THE WAR—MAKE MORE BATTERIES

BULLETIN No. 6. EDISON STORAGE BATTERY CO., ORANGE, N. J.

These ads reminded Edison factory workers of the part they played in the war effort and encouraged them to increase production.

allow Edison to ship fifty tons of carbolic acid per month. The British Foreign Office approved the request, but the British Board of Trade, which regulated merchant shipping, blocked Edison's waiver because of a shortage of ships.

Congressman Townsend wanted to know why, if the chemicals were so important, they were not manufactured in the United States. Edison explained that the United States had ample raw materials and ingenious chemists, but American companies did not have an economic incentive to produce chemicals because U.S. tariff policies allowed German manufacturers to dump their products on the American market at low prices. "If you & your friends in Congress will get together at once and pass a bill the same as the Canadians have passed to prevent 'dumping' in this market," Edison wrote to Townsend, "it will build up & make Americans dominant in the chemical trades."

Townsend appreciated the information but did not seem to understand the problem. "I do not quite 'get you' as the saying is, and during my next visit home I shall want to talk this matter over with you personally," he wrote to Edison. At this point, Edison, who scrawled the word "hopeless" across the top of Townsend's letter, may have concluded that political leaders could not solve his chemical problem.

Edison preferred to act, not talk. He learned that most American chemical manufacturers needed six to nine months to figure out how to make carbolic acid. Merck & Co.—based in Rahway, New Jersey—had assigned a chemist to the task and promised Edison shipments in three months. He could not wait that long.

Believing that he could invent a process in three weeks, Edison and his chemists began studying phenol-manufacturing techniques in September 1914. Within days they had developed a method of making synthetic phenol from other chemicals. Moving rapidly, Edison assigned three shifts of workers to design and construct a chemical plant on the grounds of his factory in Silver Lake (now Bloomfield), New Jersey. Nineteen days later, the carbolic acid plant was in operation, producing enough of the chemical to keep the phonograph factory "running without interruption."

By early January 1915, Edison was producing 3,400 pounds of carbolic acid per day. To make carbolic, however, Edison needed benzol, a liquid hydrocarbon (known today as benzene). The British had ample supplies of benzol, but shipping costs were prohibitive. "They have plenty of benzol in England, and there is no embargo on its shipment to America," Edison wrote to a Canadian steel manufacturer, "but the price is so high, the shipping facilities so bad, and the cost of drums so great, that I cannot afford to buy it."

THE FIRE OF 1914

While Edison worked on his chemical problem, a fire broke out at his West Orange factory complex on the evening of December 9, 1914. The fire began at 5:17 p.m. in the film inspection department, a one-story wood building that stored highly flammable motion picture film. The seventy-two employees of the laboratory and company fire departments were unable to bring the blaze under control, and soon the fire departments of West Orange and the nearby communities of Montclair, South Orange, and Newark sent fire trucks to assist. The firefighting was hampered by a lack of water pressure, which was boosted when West Orange temporarily connected its water main to South Orange. To prevent the laboratory complex from burning, firefighters focused on protecting buildings closest to the lab. Other buildings already ablaze were allowed to burn down.

There was only one fatality: an employee who reentered a burning building to retrieve personal items. Of the twenty-two buildings in the complex—which manufactured phonographs and records, dictating machines, and motion picture equipment—fifteen were destroyed. The value of the buildings and machinery lost was $1.5 million ($34.8 million today).

Edison wasted no time in rebuilding. The next day, workers began removing debris, while Edison and the laboratory staff began making plans to reconstruct the factory complex. His secretary William Meadowcroft reported, "Mr. Edison has been full of vim and ambition to get started up again quickly and there is no need for me to tell you that we have all been on the jump."

By the first week of January 1915, Edison had resumed production of cylinder phonograph records. Disc record manufacturing began on January 21, and by April the phonograph works had been completely rebuilt.

The charred smoldering ruins of Edison's West Orange manufacturing complex, destroyed by fire on December 9, 1914. Edison rebuilt the factory within months.

Edison benzol plant at the Dominion Iron & Steel Co., Sydney, Nova Scotia, April 1914.

Benzol, like phenol, is a by-product of coal tar or coal gas. Steel manufacturers produced these by-products when they turned coal into coke, a fuel used in steel furnaces and created via the heating of coal in enclosed ovens. A variety of chemicals could be extracted from the coal tar and coal gas driven off during this process. Edison's solution to the benzol problem: extract the chemical from coal gas by building absorption plants near the coke ovens of iron and steel companies.

On December 18, 1914, Edison asked the Dominion Iron & Steel Co. in Sydney, Nova Scotia, if it would allow him to build a benzol absorption plant near its coke oven. He sent the same proposal to the Cambria Steel Co. in Johnstown, Pennsylvania, on December 23.

Edison proposed a three-year contract, under which he would design, construct, and operate absorption plants at his own expense. The steel companies would permit Edison to absorb all of the liquid hydrocarbons produced in the coke ovens (up to 1,800 gallons per day) and sell him the steam he needed to operate the plant. In exchange, Edison agreed to pay eighteen cents per gallon for the benzol he extracted. Edison also agreed to purchase other coke oven by-products, including tuloul (now known as toluene), xylol (today called xylene), and solvent naptha. Edison would then sell these chemicals, used as industrial solvents, to other manufacturers.

The Cambria Steel Co. accepted Edison's proposal on January 12, 1915, and the Dominion Iron & Steel Co. signed a contract with Edison a month later. On February 22, he completed construction of his absorption plant at Johnstown, and Edison's benzol plant in Nova Scotia began operating on April 12. In early March, the inventor formed a partnership with Mitsui & Co., a Japanese industrial conglomerate, or *zaibatsu*, to construct a third benzol recovery plant in Woodward, Alabama. The Alabama plant, constructed at the Woodward Iron Co., began operating on May 31.

EDISON'S VIEWS ON MILITARY PREPAREDNESS attracted the attention of U.S. Secretary of the Navy Josephus Daniels. In July 1915, Daniels asked Edison to head an advisory board to evaluate technical ideas submitted to the navy by the public. Edison agreed, provided that he would not have to handle administrative matters and would be free to pursue his own war-related research. Edison's chief engineer, Miller Reese Hutchison, joined him on the board.

Eleven technical and scientific societies, including the American Chemical Society, the American Institute of Mining Engineers, the American Aeronautical Society, and the American Institute of Electrical Engineers, selected the other board members. These societies named several prominent engineers and inventors, including General Electric research director Willis R. Whitney; Leo H. Baekeland, a chemist who invented an early plastic called Bakelite; gyroscope inventor Elmer Sperry; and Frank J. Sprague, an electric railroad inventor who had worked for Edison in the 1880s.

Edison with members of the Naval Consulting Board on the east steps (facing the White House) of the State, War and Navy Building, October 7, 1915. Navy Secretary Josephus Daniels stands to Edison's left, and Assistant Navy Secretary Franklin D. Roosevelt is in the first row, far left.

The Naval Consulting Board met for the first time on October 7, 1915, adopting operating rules and creating fifteen subcommittees organized by subject, including submarines, ordnance and explosives, mines and torpedoes, and ship construction. These subcommittees evaluated invention proposals submitted by the public.

The board received approximately 11,000 suggestions, most dealing with the submarine. Of these submissions, only 110 were referred to one of the subcommittees for evaluation, while only one—the Ruggles Orientator—was produced during the war. The orientator, a simulated pilot's seat mounted on gimbals, allowed instructors to

LEFT: On January 19, 1918, the Naval Consulting Board appropriated funds to construct a model of William Guy Ruggles's orientator, a flight simulator. Ruggles built several orientators for the U.S. Army Air Service. RIGHT: American inventor Peter Cooper Hewitt, pictured in this early-twentieth-century photograph, worked with Elmer Sperry to develop the pilotless Hewitt-Sperry Automatic Airplane as part of the Naval Consulting Board.

simulate the aircraft motion for pilot trainees. The orientator also allowed the military to study the physiological effects of different flight conditions on pilots.

In February 1917, the board created a Special Problems Committee to address the issue of protecting ships from submarine attack. This committee received thousands of suggestions from the public for shields and nets to protect surface vessels from submarines. Other ideas for camouflage and smoke reduction (to reduce the visibility of ships) were also tested. None of these ideas, however, were practical.

The technical research conducted individually by board members was more consequential. Elmer Sperry developed a number of improvements for airplanes and submarines, including a device to detect hydrogen in submarines, improved steel airplane propellers, and remote-control devices for aerial bombs. Hudson Maxim invented improved contact mines and torpedo fuel. Peter Cooper Hewitt experimented on helicopters, while Frank Sprague developed depth charges, underwater fuses, and armor-piercing shells.

Edison began researching war-related technical problems in January 1917. He outfitted a laboratory in a vacant casino on Eagle Rock Mountain—near the West Orange laboratory—to test equipment for locating gun positions by sound. In the spring he conducted experiments at the Sandy Hook, New Jersey, naval station. Between August and October, he spent six weeks on Long Island Sound, experimenting on the USS *Sachem*, a 186-foot private yacht that the navy had acquired for Edison's research.

Most of Edison's naval research in 1917 focused on protecting surface ships from submarine attack. He studied methods of camouflaging ships and recommended that cargo ships burn anthracite coal, which would lessen smoke emissions. He also investigated ways to quickly turn ships under torpedo attack and equipped the *Sachem* with electrical instruments to detect submarines by sight, sound, and magnetic field.

Edison learned that previous submarine detection research had been conducted on stopped ships that had all their machinery shut down. He did not think this was practical. "Of course in trying to save a cargo boat, I could not stop her. I had to hear her going at full speed. All experiments were conducted when seas were high and at 10 1/2 knots." To detect submarines under normal operating conditions, Edison attached listening devices to a ship's bowsprit. As he recalled, "I had discovered, while you could get good listening in undisturbed water, it was a different proposition when you had waves or another boat disturbing you . . . finally we got her so it would run in any sea."

In the fall of 1917, Edison moved his war research to Washington, D.C. From October 1917 to January 1918 he used an office in the State, War, and Navy Building near the White House (now called the Eisenhower Executive Office Building) to collect data on Allied shipping losses. At the conclusion of this research, Edison discovered that the United States and its Allies were using prewar shipping

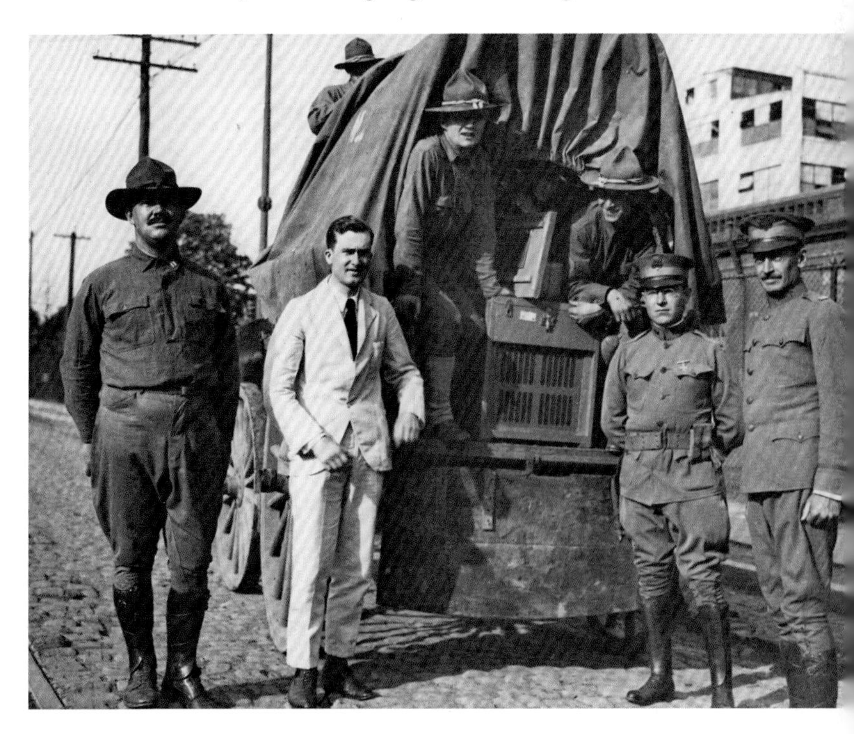

RIGHT: Charles Edison with U.S. Army soldiers and the Army and Navy phonograph. To entertain soldiers in the field, Edison sold a portable disc phonograph to the U.S. government and service organizations like the Red Cross.

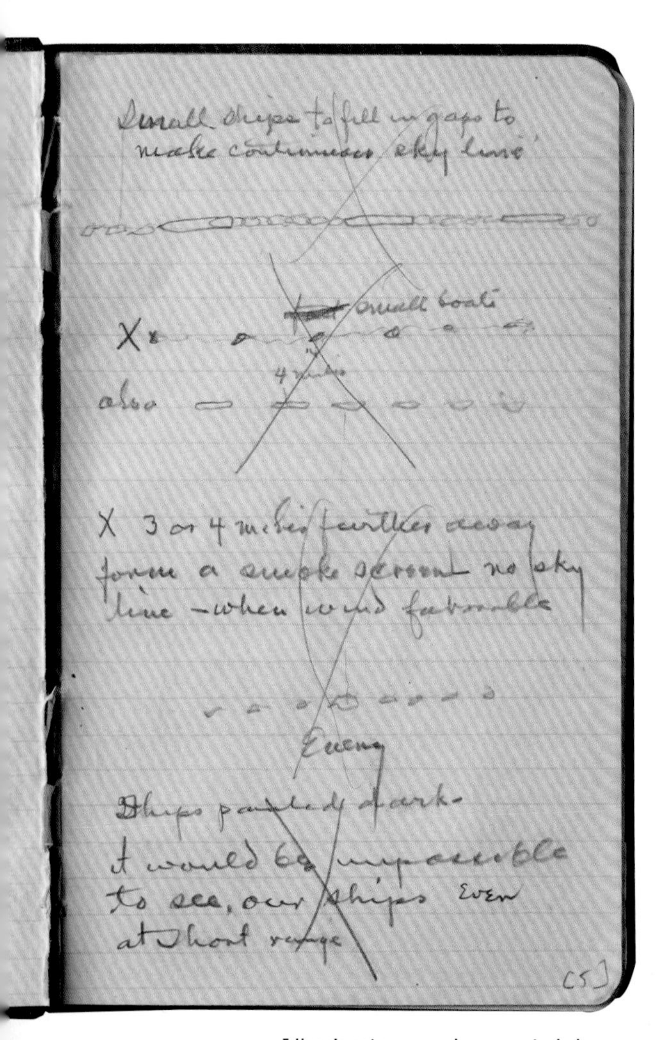

Edison's notes on smoke screen techniques designed to protect surface ships from submarine attack.

routes, which made it easier for enemy submarines to target ships, and that many ships were passing through danger zones during daylight. Edison recommended that the allies change their shipping schedules to avoid German submarines.

AT ITS FIRST MEETING on October 7, 1915, the Naval Consulting Board adopted Edison's proposal for a permanent naval research laboratory. The proposal called for a laboratory constructed at a tidewater location deep enough to dock a battleship and near—but not in—a large city. Proximity to a large urban area would give the lab easy access to supplies and labor. As the *New York Times* reported, "The laboratory should be of complete equipment to enable working models to be made and tested . . . there should be a pattern shop, a brass foundry, a cast iron and cast steel foundry, machine shops for large and small work." Managed by civilians rather than naval officers, the laboratory would also include a chemical lab, an optical grinding department, a motion picture department, and drafting rooms.

Edison wanted a well-equipped laboratory, capable of producing new inventions rapidly. He wrote Benjamin Tillman, a U.S. senator from South Carolina, "For quick action, it is necessary to have a considerable stock of supplies of every kind in order that there shall be no delay when the naval authorities want results quickly." Edison considered the proposed laboratory vital to national defense. "If the USA desires that the fighting machinery be kept up to date & not excelled or even approached by that of any other nation . . . then the experimental laboratory is the only possible way this object can be obtained." The board recommended an appropriation of $5 million ($115 million today) to construct and equip the laboratory, with an annual operating budget of $2.5 to $3 to three million ($57.7 to $69.3 million today).

In 1916, Congress appropriated $1 million ($21.1 million today) for the new laboratory—considerably less than the recommended $5 million. A committee of the Naval Consulting Board subsequently revised the first proposal. Instead of a fully equipped facility to design and construct full-scale prototypes, the committee recommended a scaled-back laboratory under the supervision of naval officers that would focus on basic scientific research and testing. The committee also recommended Annapolis, Maryland, near the U.S. Naval Academy, as the site of the laboratory.

Edison disagreed with the committee's report. He preferred Sandy Hook, New Jersey, as the site for the laboratory. More significantly, he objected to the lab's focus on pure scientific research and continued to advocate a fully equipped facility capable of producing practical inventions. Because of the war, construction of the naval research laboratory in Washington, D.C., did not begin until 1920. When it opened in 1923, operations focused on radio and sound detection research.

Edison aboard the USS *E-2* to inspect his storage batteries, December 1915. Launched on June 15, 1911, the *E-2* patrolled for German submarines off the U.S. East Coast during the war.

Lloyd N. Scott—the naval officer who wrote the Naval Consulting Board's official history in 1920—framed Edison's objections to the new laboratory plan as a dispute between his old-fashioned, nineteenth-century approach to industrial research and younger, university-trained engineers and scientists, who placed a higher value on scientific research. While Edison was not opposed to pure scientific research generally, he believed that the vast sums of money the government and industry had spent on this type of research had yielded few practical results and that a laboratory devoted exclusively to scientific investigation would be ill prepared to deal with the navy's technological problems.

Underlying this perspective was Edison's view that the navy was not an innovative organization. In 1918 he jotted in a pocket notebook, perhaps while at sea as he worked on the navy's problems, "Nobody in Naval will do anything on account of the risks that

Edison with Navy secretary Josephus Daniels aboard the battleship USS *New York* at the Brooklyn Navy Yard, October 12, 1914. The fourth U.S. Navy ship named after the state of New York, she was stationed on blockade duty in the North Sea during World War I.

an innovation will bring in . . . no training at Annapolis to cultivate the imagination." For Edison, the debate over whether the Navy Research Laboratory should pursue scientific or practical research was less important than whether it would have all the resources it needed to address the navy's technological problems. Just as important was Edison's belief that the navy needed to create an environment that encouraged innovation and risk taking.

"THE EUROPEAN WAR HAS SERVED TO DRAW ATTENTION TO THE FACT THAT MANY AMERICAN IDEAS AND INVENTIONS HAVE BEEN ALLOWED TO SLIP BY . . ."

THE NAVAL RESEARCH LABORATORY is significant because it was the federal government's first permanent facility devoted to producing new technological innovations. On a limited basis, the U.S. government had promoted firearms research at its Springfield, Massachusetts, and Harpers Ferry, West Virginia, armories since the early nineteenth century. During the Civil War, Abraham Lincoln created the National Academy of Sciences to advise federal agencies on scientific and technical issues, and the federal government created a few small laboratories in the late nineteenth century that only evaluated private-sector research.

Following the opening of the Naval Research Laboratory in 1923, the U.S. government assumed a greater role in financing technical research. A number of significant technological advances in the twentieth century—including radar, jet engines, atomic energy, digital computers, improvements in vacuum tubes, Global Positioning Systems (GPS), and the Internet—were the result of federal support. The U.S. government assumed the risk of developing many of these technologies to improve the nation's defense capabilities.

That is the legacy of Edison's role during the First World War. By advocating industrial preparedness and promoting the Naval Research Laboratory, he encouraged the view that technical research and innovation were vital to national security. Edison's awareness of the relationship between innovation and security influenced his last major research project in the 1920s: a search for a domestic source of natural rubber.

PORTLAND
CEMENT
NEWVILLAGE, N.J.
6-30

12

RUBBER

"I AM TRYING TO ASSURE A DOMESTIC SUPPLY OF RUBBER FOR THE UNITED STATES—IF THE TIME SHOULD COME WHEN IT IS NEEDED."

WITH THE FINANCIAL SUPPORT of his friends—automobile maker Henry Ford and tire manufacturer Harvey Firestone—Edison organized the Edison Botanic Research Corporation in 1927 to fund rubber experiments at West Orange and Fort Myers. From 1927 to 1929, he collected and tested the rubber content of thousands of domestic plants. In the last two years of the project, 1930 and 1931, Edison focused on selectively crossbreeding the most promising plant varieties.

Natural rubber is a hydrocarbon polymer found in plant latex. In its natural state, rubber is too soft and unstable for practical use. Vulcanization—a process, invented in the 1830s, of treating natural rubber with heat and chemicals—made it a stable industrial raw material.

In the nineteenth century, the best source of rubber was a tree native to the Amazon River basin, *Hevea brasiliensis*. Brazil controlled the world's rubber supply until 1876, when a British planter smuggled *Hevea* seeds out of Brazil and planted them in Malaysia. By 1900 the British controlled vast rubber plantations in Malaysia and their other Southeast Asian colonies, and the Dutch had established plantations in Indonesia.

PAGES 204–205: Interior of Edison's botanic research lab in Fort Myers, Florida, as it appears today. BELOW: Edison with his camping companions Henry Ford, naturalist John Burroughs, and Harvey Firestone at Yama Farms Inn, Napanoch, New York, on November 15, 1920.

Rubber consumption soared in the United States in the early twentieth century, as demand for automobile and bicycle tires, electric wire insulation, shoes, industrial tubing, and other products increased. By the 1920s, Americans consumed more than 70 percent of the world's rubber output—but produced none of it.

During the 1920s, political and industrial leaders were concerned that war, political crisis, or natural disaster would disrupt American access to rubber grown on British- and Dutch-controlled plantations in Asia. The Stevenson Plan, enacted by the British government in 1922 to stabilize rubber prices by restricting output, only increased American fears of rubber shortages.

Edison and Harvey Firestone watch M. A. Cheek tap a rubber tree at Fort Myers, March 1925. Cheek was a rubber expert employed by Firestone.

Edison was aware of the rubber problem. Rubber was an important component in his storage battery. During the First World War, he told a reporter that some varieties of American milkweed might produce rubber. He even grew rubber plants in the Glenmont greenhouse. In 1921 he considered making cylinder phonograph records out of rubber instead of the more expensive celluloid and, in 1925, constructed a factory to manufacture hard rubber for storage batteries.

Edison probably discussed the rubber problem during his camping trips with Ford and Firestone, who were heavy users of rubber. Firestone noted after their 1919 camping trip that Edison was remarkably well informed about the industrial elastomer.

As rubber prices increased in the early 1920s, Firestone and Ford became more interested in finding a domestic source. Firestone adopted the slogan "America Should Grow Its Own Rubber" and lobbied the federal government to survey the world's rubber supply and support domestic rubber research. Ford and Firestone also began supplying

Edison with plant samples, seeds, and scientific literature on rubber. Edison conducted his first rubber experiments in 1923, when he investigated the possibility of extracting rubber from milkweed. Edison's chemists also began experiments on chemical rubber extraction processes. In 1924, Edison received a supply of Brazilian rubber tree seeds and planted them at Fort Myers.

When he began rubber research in earnest in 1927, Edison had a clear conception of the type of plant he wanted: a fast-growing perennial weed with broad leaves that could be harvested mechanically. To expand the area of rubber cultivation north of Florida, he also sought a frost-tolerant plant. Edison explained his objective in a December 1927 *Popular Science Monthly* interview:

> We are looking for an annual crop, something which a farmer can grow in the field, by machinery, which will come to maturity in eight or nine months, which can then be harvested by machinery by processes almost entirely mechanical, with the least amount of hand labor. It must be something which will stand light frosts, for there is no part of the United States where there are not occasional frosts.

Beyond the botanical specifications, Edison carefully estimated the cost of growing, harvesting, and processing a plant crop to supply the 450,000 tons of rubber that the United States consumed annually. From these calculations, he found that he could not produce domestic rubber for less cost than importing rubber from Southeast Asia. As a result, Edison regarded the rubber project as an emergency measure, to be implemented when war or international crisis justified the additional cost of growing domestic rubber.

Although Edison was already familiar with the rubber literature in 1927, he read more extensively on the subject, including abstracts of foreign-language articles provided by an assistant, Barukh Jonas. He needed to know more about what was going on inside plants, so he also studied plant biology. In July 1927, he visited the library of the New York Botanical Library to study tropical rubber plants.

During the spring of 1927, Edison and his chemists developed a process to extract and test the rubber content of plants. Plant leaves were dried and crushed in a meat grinder, then boiled for two hours. After boiling, the leaves were dried and crushed again, then sifted through a mesh screen. Edison used acetone in a Soxhlet extractor—a laboratory apparatus that separated soluble compounds from liquids—to dissolve any nonlatex materials from the crushed plant. After filtering out the acetone and dissolved material, he dried the leaves again. He then added benzene to the leaves to dissolve the latex, which remained after the benzene evaporated. This process could take up to eight hours.

In the summer of 1927, Edison began collecting samples of lactiferous—or milk-producing—plants. He cast a wide net, sending fifteen collectors to Cuba, Puerto Rico, and states along the Gulf of Mexico and the Atlantic seaboard. Their instructions were simple: "Cut everything in sight." The collectors kept detailed records, noting plant location, size, and soil condition, then packed, labeled, and shipped the samples to West Orange and Fort Myers for testing. Edison also received plant samples from private collectors and Union Pacific Railroad station agents. To determine the plant's rubber content, Edison's chemists weighed samples on sensitive scales or balances, testing more than 1,000 plant specimens in 1927. During the spring of 1928, the inventor supplemented his extensive plant collection with an additional 2,000 samples that he had personally gathered throughout southern Florida, including the Everglades and the Lake Okeechobee area.

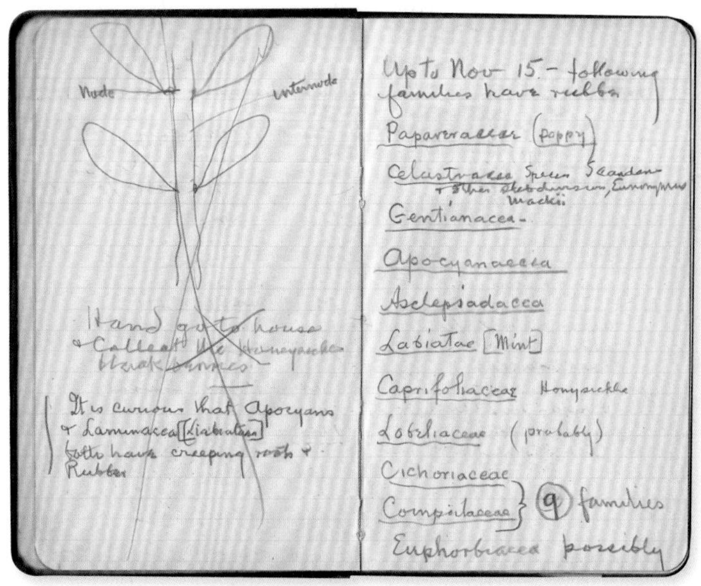

TOP: Edison used this distiller in the West Orange chemistry lab to recover acetone and benzene after they were used to extract rubber. BOTTOM: During the rubber project, Edison studied the physical characteristics of many plant varieties. In this 1929 list, he identified some of the plant families containing latex.

EDICRAFT APPLIANCES

In a 1912 *Good Housekeeping* magazine interview, Edison predicted that "the kitchen of the future will be all electric, and the electric kitchen will be as comfortable as any room in the house." Edison's vision became reality as more American homes became electrified in the early twentieth century and manufacturers began marketing vacuum cleaners, toasters, refrigerators, and other electric appliances.

As phonograph sales steadily declined in the late 1920s, Thomas A. Edison, Inc., sought other products to manufacture in West Orange to keep its factory in operation. In 1928 the company introduced a line of electric appliances—coffeemakers, waffle bakers, sandwich grills, electric irons, vacuum cleaners, and aquarium heaters—that it called the Edicraft Products.

The coffeemaker, named the Siphonator, came in three models priced from $17.50 for the least expensive to $87.50 for the Menlo to $125 for the Manhattan ($230, $1,150, and $1,640 today, respectively). The Menlo and Manhattan models came with serving trays, sugar bowls, and creamers. The waffle baker retailed for $18 and the toaster, for $15.

The Edicraft products reflected the West Orange laboratory's shift from innovation to product engineering in the 1920s. Laboratory researchers were no longer developing new inventions; they were engineering existing technologies to meet the company's manufacturing and marketing needs. The Edison Co. purchased the patent rights for the Edicraft line's most notable technical feature, a durable thermostat called the Birka regulator, which had been invented in Europe.

Edison was not personally involved with the Edicraft line, but his emphasis on high quality and technical performance influenced their design. The Edicraft products would incorporate "good electrical performance, good mechanical performance and an external appearance which would appeal to the buyer and reflect the engineering design within the device." In other words, performance and attractiveness—not price—were the selling points.

To promote the appliances, the company used Edison's name, image, and reputation. Edicraft advertisements featured Edison's signature and his personal message: "Edicraft products are the only electrical appliances developed in my laboratories, made in my factories and authorized to carry my signature." The company employed what marketing experts today call "brand signaling," that is, using Edison's reputation as an inventor to convey the desirable qualities of

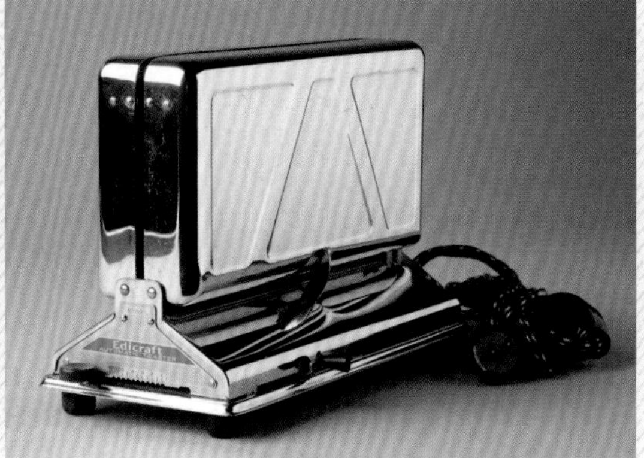

the Edicraft products. As one ad promised, "The signature of Thomas A. Edison assures you of its mechanical and electrical correctness."

The company marketed the Edicraft line to upper-middle-class households—families that could afford to buy more expensive appliances but were not wealthy enough to employ servants. The coffeemaker, sandwich grill, and waffle baker offered what the company called "Table Cookery": "For all those times when it is really a nuisance to go out into the kitchen and prepare a meal, table cookery comes graciously to the housewife's aid." To help consumers plan meals, the company issued a booklet with recipes for macaroni waffles, peanut butter omelets, and broiled bananas and bacon.

The Edison Co. introduced the appliances at the beginning of the Depression in the 1930s, when most consumers were sensitive to high-priced products. At a time when the average office worker earned $32.50 per week ($438 today) and competing models sold for $2.50 ($33 today), most could not afford to spend $18 on a waffle baker. The company stopped actively marketing the Edicraft appliances in 1932 and stopped production in 1935.

ABOVE, LEFT TO RIGHT: Edicraft electric iron, toaster, and coffeemaker. LEFT: Edicraft ads emphasized Edison's reputation for reliability and durability, but the appliances were priced beyond the reach of most consumers in the early 1930s.

At the end of the summer, Edison reported promising results to Ford and Firestone. He told Firestone, "Experiments in the chemical end are progressing fairly well," and noted in another letter, "Things look very promising for us, and with the cooperation we are getting we certainly will have rubber Trees galore in Florida." One Edison researcher had identified rubber in 26 percent of the plants he tested. Edison himself estimated that there were more than 38,000 species of plants that contained some rubber.

IN THE FALL OF 1927, Edison sent an assistant, William Benney, to Fort Myers to supervise repairs to the old 1886 laboratory. He planned to use the laboratory for rubber research when he traveled to Florida in January for his annual vacation. The building, which had been used to store surplus furniture, was in disrepair. Edison used the old lab until June, when workers dismantled the building piece by piece and loaded it on a train bound for Dearborn, Michigan, where Henry Ford planned to reconstruct it at Greenfield Village.

As the old laboratory left for Michigan, workers completed construction of a new botanic research laboratory. It included a plant nursery, storage barn, gas-powered oven for drying plants, and Portland cement vault used to store documents and supplies. There was

Edison examining a goldenrod sample at Fort Myers, 1931.

also a machine shop with a lathe, grinder, drill press, and other tools for cutting and shaping metal; a photographer's dark room; a grinding room for crushing plants; and space for a glassblower to make specialized glassware.

Edison's rubber research team included eight chemists led by Francis S. Schimerka, plant collectors, and two machinists (Fred Ott and Joe Ziemba). Jerome Osborne managed the corporation's records, herbarium, and reference library. William Benny was the superintendent of the Fort Myers laboratory. Walter Archer worked as a garden laborer in Florida, and Barukh Jonas was employed as a researcher. Edison also employed botanists and a glassblower.

After investigating 17,000 plants and identifying several with some promise—including oleander, flame vine, and black mangrove—in May 1929, Edison concluded that goldenrod, a wildflower that grew abundantly in southern Florida, was *the* plant. Its rubber yield was not as high as that of some other plants, but goldenrod had a couple of advantages going for it: it grew quickly, and its latex could be easily extracted by Edison's chemical process. Edison also believed that mechanical harvesting methods could be developed for goldenrod, which

Edison botanic research lab, Fort Myers, Florida.

would help lower labor costs, and that selective breeding could increase goldenrod's rubber yield. For this research, he set aside nine acres of land on the grounds of the Fort Myers laboratory to grow varieties of goldenrod.

Goldenrod, however, turned out to have several limitations that prevented its adoption as a domestic rubber source. It was difficult to fertilize and did not grow well north of Florida. Edison was also unable to design an efficient mechanical harvester for it.

BY 1931, THERE WAS a rubber surplus, and the conditions that prompted his rubber experiments had changed. The widespread unemployment and reduced industrial production caused by the Depression resulted in lower demand for rubber, and, consequently, its price dropped. Nevertheless, after Edison's death in October 1931, Mina Edison decided that the rubber research should continue under the supervision of her brother, John V. Miller, who, along with a team of chemists, aimed to improve the rubber yield from goldenrod. In 1934 the Edison Botanic Research Corporation, faced with mounting research costs in an economic depression, considered whether it should stop the project or convince the federal government to assume control of the experiments.

Harvey Firestone and his son, Roger, with Edison in his Fort Myers botanic research lab, March 15, 1931.

In early April, Charles Edison and John Miller met with Secretary of Agriculture Henry A. Wallace, who was interested in the research but explained that the government had cut its budget for rubber experiments. Later that month, the U.S. Department of Agriculture agreed to transfer parts of the goldenrod project to its experimental station in Savannah, Georgia.

Lack of funds, however, hindered federal research, and in January 1935, Secretary Wallace asked President Franklin D. Roosevelt to increase funding. Roosevelt did not consider the rubber project as important as other needs and declined to request additional money from Congress. Mina Edison, who regarded the rubber project as an important part of her late husband's legacy, supported research at Fort Myers until May 1936, when she agreed to dissolve the Edison Botanic Research Corporation.

Edison's views on the strategic value of domestic rubber were confirmed during the early years of the Second World War, as the Japanese seized control of rubber plantations in Southeast Asia. President Roosevelt appointed a committee that included physicist and MIT president Karl Compton, chemist and Harvard University president James Conant, and prominent financier Bernard Baruch to study the problem. The committee studied earlier efforts, including Edison's research and later attempts to develop synthetic rubber. Ultimately, however, the committee decided that synthetic materials—not domestically grown latex-bearing plants—offered the best solution to meeting the nation's rubber needs.

Ford, Edison, and Firestone outside the botanic research lab in Fort Myers, sharing a laugh.

13

REMEMBERING THE WIZARD

"IF I HAVE SPURRED MEN TO GREATER EFFORTS, AND IF OUR WORK HAS WIDENED THE HORIZON OF MAN'S UNDERSTANDING EVEN A LITTLE, AND GIVEN A MEASURE OF HAPPINESS IN THE WORLD, I AM CONTENT."

THOMAS AND MINA LEFT New Jersey for what would be his last visit to Fort Myers on January 20, 1931. On his eighty-fourth birthday, Edison dedicated a two-lane drawbridge, named in his honor, over the Caloosahatchee River. The Edisons celebrated their forty-fifth wedding anniversary on February 24. In March, Henry Ford joined them for a visit to Harvey Firestone in Miami Beach, and in April a group of local fifth graders toured the botanic research laboratory. Edison lunched on milk and crackers on the banks of the Caloosahatchee with schoolgirls from the Sarasota Open Air School.

Meanwhile, Edison continued his rubber experiments. As he told the Butte, Montana, *Standard*, his rubber research was "coming along nicely but slowly." He expected to live to 100 and, by that time, see rubber plantations in the United States. Edison discussed plans with Henry Ford to mount a rubber exhibit at the 1933 Chicago Century of Progress Exposition. In late May he spilled acid on his hands during a rubber experiment.

Edison returned to West Orange on June 16. Excessive heat and fatigue from the trip prevented his immediate return to the laboratory and forced him to rest at Glenmont. On July 24, Mina's brother, John Miller, publically denied newspaper reports that Edison had retired but admitted that the aging inventor had visited the laboratory only once since his return from Florida.

Edison collapsed at Glenmont on August 1. While newspapers reported that he was near death, the family summoned his doctor, Hubert S. Howe, who was vacationing on Long Island's north shore. The family began issuing regular updates on his condition to the press, while the police assigned a protection detail to keep a curious public away from Glenmont. The Peoria, Illinois, *Star* editorialized, "Nature is about to collect a debt due from the greatest inventor the world has ever known. . . . [Edison] will never be able to work again and this fact, if nothing else may speedily end the old man's days."

Edison, however, rallied, and by August 4, Howe reported, "Mr. Edison slept eight hours and had the best night so far. He ate his breakfast with relish, read the morning papers and showed evidence of returning strength and health." The family stopped the daily press briefings, and the police reduced the guard from five officers to two.

Edison was well enough to take a short drive through the Orange Mountains and enjoy a dinner of tomatoes, peas, fruit, and milk. Forbidden to smoke cigars, he told his doctor, "If I live through my 84th year, I'll probably live ten years longer." Dr. Howe tactfully agreed, but

PAGES 216–217: A crowd gathered in front of West Orange, New Jersey, town hall for the Edison Pageant of Progress, May 1940.

he knew better. Edison's heart and pulse were strong, but he suffered from diabetes, Bright's disease (of the kidneys), stomach ulcers, and uremic poisoning (symptomatic of kidney failure). As Howe told one reporter, "In my opinion he never will be strong enough to work. He will never be out of danger."

By the end of September, it was clear to the family, friends, doctors, and the press that the end was near. Edison slept comfortably, but the lack of improvement depressed him. By October 13, he was refusing fluids, eating very little, and recognizing no one except Mina. The family resumed daily briefings for the reporters camped on the first floor of Glenmont's two-story garage. One of Edison's last visitors was "Uncle Bob" Sherwood, a retired circus clown and old family friend who later published a memoir, *Hold Yer Hosses, the Elephants Are Coming!* On October 15, Edison slipped into a coma. He died in his bedroom at Glenmont on October 18 at 3:24 am.

Crowds of mourners wait on Lakeside Ave. outside the West Orange lab to view Edison's body in the library, October 19, 1931.

James Earle Fraser, a sculptor who had worked with Augustus Saint-Gaudens and designed the Indian Head nickel in 1913, prepared a death mask and cast of Edison's hands before his body was moved to lie in state in the laboratory library. On Monday, October 19, and Tuesday, October 20, 50,000 mourners filed past Edison's casket. Employees of the Edison Company paid their respects first, followed by the public. Average men and women, schoolchildren, and the discharged passengers of limousines driven by liveried chauffeurs stood in long lines snaking from the library through the laboratory courtyard and onto Lakeside Avenue.

EDISON'S LONGEVITY INQUIRY

In October 1930, Edison answered a five-page health questionnaire for Irving Fisher, a Yale University economist, vegetarian, and exercise advocate who had coauthored *How to Live: Rules for Healthful Living Based on Modern Science* (1915). The questionnaire—which asked specific questions about the respondent's medical history, diet, and sleeping habits—reveals the state of Edison's health in the last year of his life.

For an eighty-three-year-old man who no longer had his own teeth, never exercised, bathed less than once a week, chewed tobacco continually, smoked two or three cigars a day, and suffered from indigestion and gas, Edison was in pretty good health. At five feet nine and a half inches, Edison weighed 170 pounds. His heart rate was "about 74," and his blood pressure was normal for his age.

Genetics may have contributed to his longevity. His mother died at age seventy of what Edison called "worry," while his father lived to age ninety-four. Edison did not know how long his father's parents had lived, but his maternal grandfather reached the age of 103, and his maternal grandmother died at ninety.

Because he had no teeth, Edison claimed, he confined his diet to orange juice and six glasses of milk per day. Following the teachings of Luigi Cornaro, a fifteenth-century Venetian who lived to age ninety-eight on a restricted-calorie diet, Edison ate moderately for most of his life. In

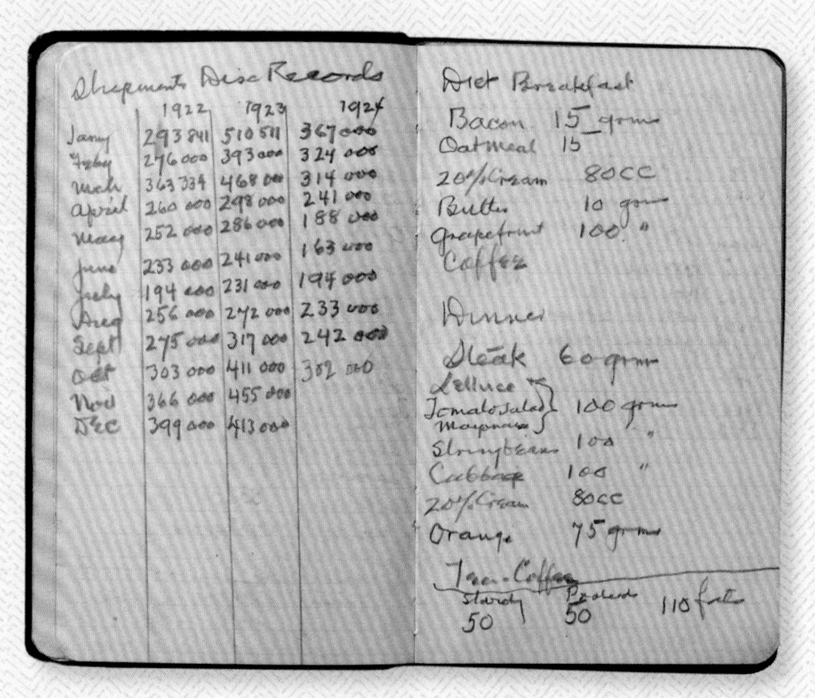
Shipments Disc Records

	1922	1923	1924
Jany	293 841	510 511	367 000
Feby	276 000	393 000	324 000
Mch	363 334	468 000	314 000
April	260 000	298 000	241 000
May	252 000	286 000	188 000
June	233 000	241 000	163 000
July	194 000	231 000	194 000
Aug	256 000	272 000	233 000
Sept	275 000	317 000	242 000
Oct	303 000	411 000	382 000
Nov	366 000	455 000	
Dec	399 000	413 000	

Diet Breakfast
Bacon 15 grm
Oatmeal 15
20% Cream 80CC
Butter 10 grm
Grapefruit 100 "
Coffee

Dinner
Steak 60 grm
Lettuce
Tomato Salad 100 grm
Mayonnaise
Stringbeans 100 "
Cabbage 100 "
20% Cream 80CC
Orange 75 grm

Tea - Coffee
Starch 50 Protein 50 110 fat

Mixing personal and business affairs, Edison tracked the shipment of phonographs and his daily food consumption in this 1924 pocket notebook.

response to one of the many letters he received asking about his eating habits, Edison said he "followed no special diet, I eat every kind of food, but in very small quantities, 4 to 6 oz to a meal." In 1921 he listed the following meal plan: For breakfast he drank a cup of coffee (half milk, half coffee) and ate two pieces of toast, plus another piece of toast with two small sardines. Lunch was a glass of milk with two pieces of dry toast. For dinner, Edison drank two glasses of milk and ate three pieces of dry, thin toast, a small piece of steak, and a small baked potato, followed by one piece of chocolate with nuts.

Edison did not follow Fletcherism, a popular food trend named after Horace Fletcher, a nineteenth-century health-food promoter who believed food should be chewed thirty-two times before it was swallowed. When the *New York Times* asked Edison in 1914 if he Fletcherized, he replied, "Fletcherize nothing, bolt I bolt my food; that's the thing. Fletcherized food is too quickly digested. All animals bolt their food."

Edison also had strong views on sleep. In Fisher's survey, he noted that he slept with his windows open for about six hours a night—an increase from the less than five hours he typically slept throughout his life. "Eating too much is a habit, just like sleeping too much," Edison argued. "If the sun never set men would get out of the habit of sleeping, they would get used to going without." He believed sleep was an evolutionary holdover from early human history, when there was little to do after sunset, and predicted, "The man of the future will spend far less time in bed than the man of the present does, just as the man of the present spends far less time in bed than the man of the past did." Of course, the technologies he developed—like the electric light, phonograph, and motion pictures, which extended day into night—may also have contributed to sleep-deprived modern life.

In August 1911, Edison summarized his rules for a long life: sleep six hours, never retire, study music, and eat a handful of solids at each meal.

Edison on the front lawn of Glenmont, June 30, 1917.

The casket was moved back to Glenmont for a private funeral on October 21, attended by the family and 400 close friends and prominent mourners, including the Fords, the Firestones, First Lady Lou Henry Hoover, and General Electric president Owen D. Young. Rev. Stephen J. Heben read passages from the Bible, and Phillips Exeter Academy headmaster Lewis Miller delivered the eulogy. A performance of "I'll Take You Home Again, Kathleen" and "My Little Grey Home in the West" by organist Alexander Russell and violinist Arthur Walsh followed. After the service, Edison was interred in nearby Rosedale Cemetery.

In the days following Edison's death, his family received sympathy messages from around the world. Benito Mussolini, Pope Pius XI, and German president Paul von Hindenburg were among the prominent world leaders expressing their condolences. Church groups, professional societies, and schoolchildren sent tributes, which Mina carefully preserved in leather slipcases. One letter of sympathy came from Gus Winkler, an Al Capone hit man who wrote to Mina, "Except [*sic*] my deepest sympathy. Death is something humanity can't prevent. One must be laid away in life's everlasting sleep. May he rest in peace."

THE CANONIZATION OF EDISON began before his death. During the late 1920s, he began receiving awards and tributes from the industries that he had helped establish, including the motion picture industry, which gave him an honorary Academy Award in October 1929. On October 20, 1928, the forty-ninth anniversary of the invention of the incandescent electric lamp, Treasury Secretary Andrew Mellon bestowed on Edison a special Congressional Gold Medal. In a short radio speech delivered on that occasion, President Calvin Coolidge described Edison as "the master of applied science" who "literally brought light to the dark places of the world."

These awards were part of a larger effort to preserve an account of Edison's massive contributions to society. In the early 1920s, the Association of Edison Illuminating

The inscription on Edison's Congressional Gold Medal reads, "He illuminated the path of progress by his inventions."

Treasury Secretary Andrew W. Mellon presents Edison with the Congressional Gold Medal at West Orange on October 20, 1928.

Companies (an electric utility trade group) and the Edison Pioneers (a fraternal organization of Edison associates from the Menlo Park period) began assembling a collection of Edison memorabilia that they planned to exhibit in the United Engineering Societies building in New York. The goal, according to F. A. Wardlaw, the secretary of the Association of Illuminating Companies, was "to place all the original examples of Mr. Edison's marvelous genius possible of attainment, not only to protect that which still exists from any possibility of loss or damage, but also as a well-mounted tribute of affection and appreciation of one who has done so much to enhance the scientific and industrial progress of his native land."

Edison allowed Wardlaw to inventory historical relics in his laboratory and, in the spring of 1922, approved Wardlaw's request to repatriate the original tinfoil phonograph, which had been in London's South Kensington Museum (today the Victoria & Albert Museum) since 1880.

The request sparked a four-year dispute between Edison and the South Kensington Museum. At issue was whether Edison had given or loaned the museum the tinfoil phonograph in 1880. The museum claimed that the phonograph was a gift and refused to send it

Edison in the West Orange lab with his Edison Effect lightbulb, 1919. In the early 1880s Edison observed that electrons inside an incandescent lightbulb moved from hot to cooler elements. The phenomenon, thermionic emission, became the basis of British physicist John Fleming's 1904 invention of the radio vacuum tube.

back, but Edison contended that he loaned the machine, with the understanding that it would be returned upon request. At one point in the years-long exchange of letters and legal papers, Edison's secretary threatened to go to London and remove the phonograph from the museum personally. Finally, in 1926 the U.S. State Department and the British embassy in Washington, D.C., convinced the South Kensington Museum to settle the matter. Secretary Mellon returned the tinfoil phonograph to Edison during the Congressional Gold Medal ceremony.

Preserving Edison history during the 1920s had personal, corporate, and cultural motivations. As Edison aged, his family became concerned about how history would remember him. Preserving the sites and artifacts associated with his career, the family hoped, would help ensure an accurate portrayal of Edison's story.

The preoccupation with Edison's legacy also reflected a broader interest in honoring the nation's technological heritage. Edison represented a class of inventors and engineers who had ushered in a new technological order in the decades between the Civil War and the 1920s. In the years following the First World War, several technology museums, including the New York and Chicago Museums of Science and Industry, were created to glorify this new order and the values that helped create it, including what historian Mike Wallace called "the belief in progress, reason, the arts and sciences, education and emancipation."[1] For many Americans, Edison's career embodied these changes and values.

Edison's cultural status became a valuable commercial asset for his company, which struggled in the highly competitive phonograph market in the 1920s. Consequently, the company used Edison's persona in its advertising. As one manager noted in 1927,

> It is felt and earnestly recommended that full use should be made of the name of Thomas A. Edison and what he has done for the world in the advertising of each of the industries, and that all of these industries should embody in their advertising references to the other industries, and that Thomas A. Edison should be as a bright sun over all illuminating to the fullest extent each one.

The artifacts and documents in the West Orange laboratory became resources that could be used to create a historical narrative suited to the company's marketing needs. This advertising strategy, combined with a growing public interest in Edison's life, prompted better care of artifacts and historic documents.

The cot in the West Orange library reminds visitors of the long hours Edison spent working at the lab.

Before the 1920s, Edison's letters and business records, accumulated over a long career, were stored haphazardly in wooden crates in the library and other laboratory rooms. After the First World War, Thomas A. Edison, Inc., created a Vault Services Department to consolidate inactive records and, in the early 1920s, began identifying and inventorying historically significant material. In 1928 the company created a Historical Research Department and hired a librarian, who began cataloging the notes and documents in Edison's library.

After Edison's death, work at the laboratory continued into the late 1930s, when company managers began considering a preservation plan. In a December 1938 memo to Mina, Charles, and Theodore Edison, company executive C. S. Williams lamented the condition of the laboratory. "The present situation here is really appalling," he noted.

> The Laboratory and the buildings surrounding it resemble a jackdaw's nest. Invaluable relics and models made under Mr. Edison's continuing regard are hopelessly mixed with paint pots, old cigar boxes, records, seed catalogs, ten penny nails, and a general hodge-podge that is sickening to look at.

Williams wanted the company to preserve the buildings and their contents and to hire more staff to care for the objects and plan exhibits. With family support, the company transformed the laboratory into a memorial, opening the lab as a private museum in 1948.

In 1946, the year before her death, Mina sold Glenmont to Thomas A. Edison, Inc., with the intent of preserving it as "a memorial to my dear husband and his work." In December 1955, the U.S. Secretary of the Interior designated Glenmont "Edison Home National Historic Site," and in 1959 McGraw-Edison, the successor to Thomas A. Edison, Inc., deeded the property to the National Park Service. In 1955 Thomas A. Edison, Inc., transferred ownership of the laboratory to the National Park Service. Congress designated Glenmont and the laboratory "Edison National Historic Site" on September 5, 1962.

The Edison family used the Glenmont den for informal entertaining. Decorated with souvenirs and gifts, including a desk set from German munitions maker Frederich Krupp, the room featured a large tree every Christmas.

BY THE 1920S, the original Menlo Park laboratory, long since abandoned, was dilapidated. The state of New Jersey dedicated a memorial tablet near the site on May 16, 1925, in a ceremony attended by Thomas and Mina. In the late 1920s Henry Ford took what was left of the lab and reconstructed it at Greenfield Village, a museum he built near his

Dearborn, Michigan, home. Ford stocked the reconstructed lab with original tools and equipment that he obtained from Fort Myers and West Orange. To give the reconstruction an air of authenticity, he shipped original bricks from the New Jersey site and rebuilt the lab on top of Menlo Park soil.

Greenfield Village, a collection of historic buildings that included the Wright brothers' bicycle shop and Noah Webster's home, opened on October 21, 1929—the fiftieth anniversary of the invention of the incandescent lamp. Edison, President Herbert Hoover, John D. Rockefeller, and other notables attended the celebration—Light's Golden Jubilee—which featured a reenactment of the invention of the electric light and was broadcast nationally on the radio.

In 1932, New Jersey created an Edison Park Commission to consider proposals for a memorial and "museum of light" at Menlo Park. Among the suggestions was an eleven-mile highway connecting Perth Amboy and Plainfield, New Jersey, which would pass

TOP: Reconstructed Menlo Park lab at the Henry Ford Museum, Dearborn, Michigan, as it appears today. Dedicated on October 21, 1929, the fiftieth anniversary of Edison's electric light, the restored lab is part of Greenfield Village, an open-air museum Henry Ford created in the 1920s to preserve significant buildings in the history of American agriculture, manufacturing, and transportation. BOTTOM: Edison and President Herbert Hoover at Light's Golden Jubilee, Dearborn, Michigan, October 21, 1929.

through a chain of lakes and parkland visible to motorists. Edison's daughter Madeleine endorsed a proposal for dual monuments linking Menlo Park and West Orange. The base of the Menlo Park monument would be a stylized dynamo projecting a shaft of light into the night sky. Another idea called for a circular museum building, surmounted by a 175-foot-high shaft bearing an "eternal light." On February 11, 1937, exactly ninety years after Edison's birth, the Edison Pioneers announced plans to construct a permanent 135-foot Art Deco–style cement tower at Menlo Park, topped by a fourteen-foot-high light bulb made of segmented Pyrex glass. The tower was dedicated on February 11, 1938.

Edison's parents had sold their Milan home when they moved to Port Huron in 1854, and Edison's sister, Marion Edison Page, had purchased the Edison birthplace in 1894, living there until her death in 1900. In 1906, Edison bought the property and asked his cousin, Nancy Wadsworth, to act as caretaker.

Edison paid his last visit to Milan on August 11, 1923, following President Harding's funeral. As Thomas and Mina, accompanied by the Fords and Firestones, drove down Main Street, a local band played "Yes! We Have No Bananas." The car stopped in

Edison Memorial Tower in Menlo Park, New Jersey, illuminated at night during the 1930s. The 118-foot high Art Deco–style tower stands on the site of Edison's original Menlo Park lab.

front of a country store, where the mayor gave Edison the keys to the city. Henry Ford, who was considering a run for president, overshadowed the celebration. As Ford moved among the crowd, shaking hands and noting how clean the village looked, he heard shouts of "There's our next president!"

Before leaving for the next leg of their trip—camping in the woods of Michigan—Edison visited his birthplace and was shocked to learn that his cousin still lighted the home with oil lamps and candles. In 1944 Mina Edison decided to preserve the birthplace as a memorial. It opened as a museum on the centennial of Edison's birth: February 11, 1947.

Mina Edison continued her annual visits to Fort Myers after Edison's death, with the exception of the 1943 and 1944 seasons. In the late 1930s, she considered using the Fort Myers property on the east side of McGregor Boulevard for the construction of an Edison Memorial Library. An architect hired in 1939 submitted plans that included a Mission-style library with several reading rooms, an art gallery, and a goldenrod garden. The plan, estimated to cost nearly $100,000 ($1.6 million today), called for the removal of Edison's botanic research laboratory. In 1941 Mina rejected this idea and hired another architect, who drafted a plan for a smaller library building, an arboretum, a reflecting pool, and preservation of the research lab. For unknown reasons, Mina abandoned the idea before the groundbreaking, but the library was included in a summer 1945 proposal for the creation of an Edison Research University at Fort Myers. The U.S. Office of Scientific Research and Development drafted the proposal—calling for a university that would promote independent scientific and technical research and train scientists for war service—and submitted it to President Harry S. Truman. However, the cost was prohibitive, and the plan was dropped.

Edison's winter home in Fort Myers, Florida. Each February the City of Fort Myers celebrates Edison's birth with the Festival of Light, an annual event that began in 1938 and today includes a parade, concerts, an antique car show, and an inventor's fair.

In February 1947, Mina donated the Florida estate to the city of Fort Myers. She deeded land on the east side of McGregor Boulevard on the condition that it become a public park dedicated to Edison's memory. The property on the west side of McGregor Boulevard, along with Mina's gift of $50,000, went to a nonprofit corporation that would preserve the site as an educational and cultural resource. Upon Mina's death in August 1947, the city assumed ownership and soon opened the estate to visitors. The city of Fort Myers acquired the adjacent Ford property in 1991, and the combined sites are today operated as the Edison & Ford Winter Estates.

THE PUBLIC'S ADULATION OF EDISON verged on the hagiographic. Elihu Thomson, an inventor and cofounder of the Thomson-Houston Electric Co., thought the Wizard had received too much acclaim. In a 1915 evaluation of Edison for the president of the Massachusetts Institute of Technology, Thomson wrote, "He is undoubtedly deserving of a great amount of appreciation from his fellow men, but it has sometimes in the popular estimation gone much too far, even almost to regarding him as the sole author and pioneer in electrical work."

Mrs. W. C. Lathrop of Norton, Kansas, encapsulated Thomson's view when she wrote a letter to Edison in March 1921, thanking him for all of his inventions. "I feel that it is my duty as well as privilege to tell you how much we women of the small town are indebted to you for our pleasures as well as our utmost needs." Mrs. Lathrop credited Edison with all of the electrical appliances her family enjoyed—even the products he had not invented.

> The house is lighted by electricity. I cook on a Westinghouse electric range, wash dishes in an electric dish washer. An electric fan even helps to distribute the heat over part of the house. . . . I wash clothes in an electric machine and iron on an electric mangle [rollers used to wring water from wet clothes] and an electric iron. I clean house with electric cleaners. I rest, take an electric massage and curl my hair on an electric iron. Dress in a gown sewed on a machine run by a motor. Then start the Victrola and either study Spanish for a while or listen to Kreisler and Gluck and Galli-Curci in almost heavenly strains.

If Edison noticed the reference to his competitors (the Victor Talking Machine Co. made the Victrola, and Amelita Galli-Curci was a Victor recording artist), he did not mention this in his response; he simply asked his secretary to "thank her very much."

Mrs. Lathrop—an upper middle-class, college-educated mother of four and wife of a prominent local surgeon who had entertained the governor of her state—may not have been a typical consumer of the 1920s. But her praise reveals the public's association of Edison with the technologies that were transforming modern life.

"I HAVE, BESIDE THE INVENTOR'S USUAL MAKE-UP, A BUMP OF PRACTICALITY . . . IT WAS POUNDED INTO ME BY SOME PRETTY HARD KNOCKS."

Americans appreciated Edison for his specific inventions, but in the early twentieth century his persona became synonymous with broader values of ingenuity, creativity, and practicality. Edison embodied American faith in material progress, can-do spirit, and optimism in the future. He confirmed the belief that if you had a good idea and worked hard, you would succeed, even if you came from humble origins. In an age that saw the growing influence of large public and private organizations, Edison proved that individuals still mattered.

The cultural assumption equating Edison with innovation endures today. The lightbulb over the cartoon character's head—the universal symbol for a great idea—is inspired by Edison's most notable achievement. A number of modern management books draw upon Edison's methods of collaboration and personal characteristics of creativity, imagination, and persistence to inspire modern innovators.[2]

In an episode of *The Simpsons* ("A Tree Grows in Springfield," original air date November 25, 2012) Homer Simpson becomes obsessed with his "myPad," prompting his irritated boss, Mr. Burns, to demand that he "unhand his Edison slate." The reference, a sly joke at the expense of an old man who thinks Edison is responsible for every new technology, speaks to the Wizard's continued cultural relevance.

Edison was an exceedingly practical inventor and entrepreneur who excelled at bringing together the money, tools, technical and scientific information, and skilled workers needed to operate productive research laboratories. Within the lab, he was gifted at solving technical problems. He was also a capable product engineer who could effectively lead teams to build and test invention models and design manufacturing facilities to mass produce his inventions. It also didn't hurt that he was savvy about promoting himself. Edison instinctively knew how to project the right mix of authenticity, know-how, and optimism that

Edison with his tinfoil phonograph and dictating machine in the West Orange library, December 1913.

inspired confidence in his endeavors and attracted investors.

At Menlo Park and West Orange, Edison established a close relationship between product design, manufacturing, and marketing, demonstrating that the collaboration of these different functions could result in a more innovative operation. His efforts to market his inventions were not always successful, proving that even a good idea combined with the world's best-equipped laboratory did not always ensure success. Edison did not let failure defeat him, though. If something didn't work, he tried something else until he got it right.

While the collective impact of Edison's specific inventions was significant, even more significant was his impact on the process of invention itself. Before Edison became a professional inventor in the early 1870s, the introduction of new inventions was largely the work of independent inventors or mechanics who were fortunate enough to find the financial backing to market their ideas. Machine shops created to support textile mills and other factories were an important source of technical knowledge, but these shops existed mainly to solve manufacturing problems—not to turn out new products. Few companies, if any, supported industrial research. As a result, innovation was a haphazard process contingent on inventors finding the right investors, producers, and promoters.

Edison's achievements as a telegraph inventor convinced the managers of companies like Western Union that supporting Edison's laboratory would be an effective way of developing the technologies they needed to control and expand their markets. By establishing a reputation for reliability and by creating productive research facilities, Edison proved to the capitalists who controlled these companies that permanent industrial research laboratories would be more efficient than waiting for independent inventors to approach them with their ideas. Edison thus made technological innovation a safe and reliable investment, demonstrating that corporate capitalism could create and control facilities that turn out new ideas and products on a regular basis.

Because of Edison, team-based industrial research became an important model for innovation in the late nineteenth and early twentieth centuries. In this period, a number of large corporations, including General Electric, DuPont, AT&T, Corning, and Western Electric, had established their own research laboratories. According to historian Carroll Pursell, by the First World War there were about 375 research laboratories in the United States, and by 1931 that number had increased to 1,600.[3] Companies that could not afford to operate their own labs could use the services of contract researchers like Arthur D. Little, an MIT-trained chemist who created his consulting firm in 1886. Independent inventors like aviation pioneers Orville and Wilbur Wright could still make contributions, but in the twentieth century, teams of researchers working in industrial, government, and university laboratories were the driving force in technical innovation, introducing important developments in chemicals, pharmaceuticals, and nuclear energy, among other achievements.

In recent years, some firms have scaled back their support for research and development or closed their laboratories in an effort to reduce costs, raising questions about how these companies can remain innovative in a highly competitive, rapidly changing economy. Edison's experience is relevant today because he grappled with the same problems, and we can look to his example for ways to overcome the many challenges of operating a profitable, self-sustaining enterprise. We can stand in his laboratories and see how he organized his workspace. His lab notebooks allow us to see how he translated his ideas into tangible products. His correspondence with his employees, associates, and consumers allows us to learn how he responded to questions about the best way to design, manufacture, and market his inventions. Not only did Edison develop new technologies that formed the basis of entirely new industries, but he also pioneered the movement of investor capital inside companies—a dynamic that continues today and drives much of our economic development.

NOTES

PREFACE

1. Monetary conversions in this book were calculated at MeasuringWorth.com, a website founded by Laurence H. Officer and Samuel H. Williamson, economics professors at the University of Illinois at Chicago. Because of inflation, data gaps, and other economic variables, accurately calculating monetary values over long periods of time is difficult. The conversions in this book estimate the relative modern value of significant Edison income, cost, and price figures.

CHAPTER 3

1. Surviving records in Edison's papers allow us to track design changes in Edison's inventions, but they often do not explain why he made those changes or why he abandoned an idea.

CHAPTER 5

1. The cause of Mary Edison's death remains a mystery. The death certificate did not specify a cause, and the few surviving documents relating to her health offer conflicting information. In his 1959 biography, Matthew Josephson attributed the death to typhoid fever. Robert Conot, the first biographer to make full use of the Edison archives in West Orange, claimed she died of a vague "mind-affecting disease" (Conot, *A Streak of Luck*, 1979). Conot also noted that this was a source of shame within the Edison family and that Edison told daughter Marion that her mother died from typhoid. Biographer Neil Baldwin (*Edison: Inventing the Century*, 1995) wrote that Mary succumbed to "congestion of the brain," a generic late-nineteenth-century medical term that covered a variety of ailments. This is based on a telegram an Edison associate sent on the day of Mary's death. Paul Israel (*Edison: A Life of Invention*, 1998) accurately states that the cause is unknown. In volume seven of *The Papers of Thomas A. Edison, Losses and Loyalties, April 1883–December 1884* (2011), the editors reference a *New York World* article published shortly after the funeral, which claimed that "Mary died of an accidental overdose of medicinal morphine." The editors note that Mary suffered from unspecified uterine problems and that morphine was a common treatment, but there is no evidence to verify this explanation of Mary's death.

CHAPTER 13

1. Mike Wallace, "Progress Talk: Museums of Science, Technology and Industry," in *Mickey Mouse History and Other Essays on American Memory* (Philadelphia: Temple University Press, 1996), 78.

2. Among the works in this genre are Blain McCormick, *At Work with Thomas Edison: 10 Business Lessons from America's Greatest Inventor* (Irvine, CA: Entrepreneur Press, 2001); Michael J. Gelb and Sarah Miller Caldicott, *Innovate Like Edison: The Five-Step System for Breakthrough Business Success* (New York: Penguin, 2007); Alan Axelrod, *Edison on Innovation: 102 Lessons in Creativity for Business and Beyond* (San Francisco, CA: Jossey-Bass, 2008); and Sarah Miller Caldicott, *Midnight Lunch: The 4 Phases of Team Collaboration Success from Thomas Edison's Lab* (Hoboken, NJ: John Wiley & Sons, 2013).

3. Carroll Pursell, *The Machine in America: A Social History of Technology* (Baltimore: Johns Hopkins University Press, 1995), 223.

RESOURCES

TO LEARN MORE ABOUT THOMAS EDISON, visit the seven Edison-related historic sites and museums in the United States:

Edison Birthplace Museum, Milan, Ohio
tomedison.org
Thomas Edison Depot Museum, Port Huron, Michigan
phmuseum.org
Thomas Edison House, Louisville, Kentucky
edisonhouse.org
Thomas Edison Center at Menlo Park, Edison, New Jersey
menloparkmuseum.org
Henry Ford Museum, Dearborn, Michigan
thehenryford.org
Edison & Ford Winter Estates, Fort Myers, Florida
edisonfordwinterestates.org
Thomas Edison National Historical Park, West Orange, New Jersey
nps.gov/edis

Thomas Edison National Historical Park preserves an extensive collection of Edison's personal papers, business records, laboratory notebooks, patents, advertising material, and historic photographs.

The Thomas A. Edison Papers Project at Rutgers University has published selected documents from this collection and other archives. The project website (*edison.rutgers.edu*) has a full index of published material, a chronology of important events in Edison's life, a list of his companies, and all of his 1,093 U.S. patents. The editors of the Thomas A. Edison Papers Project are also publishing a heavily annotated fifteen-volume book edition of selected documents. As of 2012, Johns Hopkins University Press has published seven volumes, covering Edison's life up to December 1884.

Of the many books and articles about Thomas Edison, the works listed below are essential for understanding his life and work. For a complete list of the literature on Edison, see the full bibliography at the Thomas A. Edison Papers Project website (*edison.rutgers.edu/fullbib.htm*).

Gene Adair, *Thomas Alva Edison: Inventing the Electric Age* (Oxford University Press, 1996); Gene Barretta, *Timeless Thomas: How Thomas Edison Changed Our Lives* (Henry Holt, 2012); William H. Meadowcroft, *The Boy's Life of Edison* (Harper & Bros, 1921); Charles E. Pederson, *Thomas Edison* (ABDO, 2007); Martin Woodside, *Thomas Edison: The Man Who Lit Up the World* (Sterling Publishing, 2007): biographies for young readers.

Neil Baldwin, *Edison: Inventing the Century* (Hyperion, 1995; University of Chicago, 2001); and Paul Israel, *Edison: A Life of Invention* (John Wiley & Sons, 1998): the standard modern biographies. Baldwin's focus on Edison's personal life and Israel's analysis of Edison's inventions and businesses complement each other.

Eileen Bowser, *The Transformation of Cinema, 1907–1915* (University of California Press, 1990); and Charles Musser, *The Emergence of Cinema: The American Screen to 1907* (Scribner, 1990): general histories of the early motion picture industry that reference Edison.

Leonard DeGraaf, *Historic Photos of Thomas Edison* (Turner Publishing Co., 2008): Edison's story through selected images from the Thomas Edison NHP historic photograph collection.

Frank L. Dyer and Thomas C. Martin, with William H. Meadowcroft, *Edison: His Life and Inventions* (2 volumes, Harper & Bros, 1910): Edison's official biography.

Mark Essig, *Edison & the Electric Chair: A Story of Light and Death* (Walker & Company, 2003); and Richard Moran, *Executioner's Current: Thomas Edison, George Westinghouse, and the Invention of the Electric Chair* (Knopf, 2002): monographs on the AC/DC controversy.

Mark R. Finlay, *Growing American Rubber: Strategic Plants and the Politics of National Security* (Rutgers University Press, 2009): Edison's rubber research.

Robert Friedel and Paul Israel, *Edison's Electric Light: Biography of an Invention* (Rutgers University Press, 1986), revised and republished as *Edison's Electric Light: The Art of Invention* (Johns Hopkins University Press, 2010): the standard account of the invention of the electric light. The 1986 edition includes more historic photographs and technical drawings from Edison's notebooks.

Thomas P. Hughes, *Networks of Power: Electrification in Western Society, 1880–1930* (Johns Hopkins University Press, 1983): the development of Edison's electric lighting system.

Paul Israel, *From Machine Shop to Industrial Laboratory: Telegraphy and the Changing Context of American Invention, 1830–1920* (Johns Hopkins University Press, 1992): an examination of the environment in which Edison began his career as a telegraph inventor.

Thomas E. Jeffrey, *From Phonographs to U-Boats: Edison and His "Insomnia Squad" in Peace and War, 1911–1919* (LexisNexis, 2008): detailed information about Edison's activities during the First World War.

Martin V. Melosi, *Thomas A. Edison and the Modernization of America* (Scott Foresman/Little, Brown Higher Education, 1990): a brief account of Edison's life.

Andre Millard, *America on Record: A History of Recorded Sound* (Cambridge University Press, 1995); and David Morton, *Off the Record: The Technology and Culture of Sound Recording in America* (Rutgers University Press, 2000): general histories of sound recording technology.

Andre Millard, *Edison and the Business of Innovation* (Johns Hopkins University Press, 1990): the influence of the machine shop culture on Edison.

Charles Musser, *Before the Nickelodeon: Edwin S. Porter and the Edison Manufacturing Company* (University of California Press, 1991); *Thomas A. Edison and His Kinetographic Motion Pictures* (Rutgers University Press, 1995); *Edison Motion Pictures, 1890–1900: An Annotated Filmography* (Smithsonian Institution Press, 1997); and Paul Spehr, *The Man Who Made Movies: W.K.L. Dickson* (John Libbey, 2008): the development of the motion picture business at West Orange.

William S. Pretzer (editor), *Working at Inventing: Thomas A. Edison and the Menlo Park Experience* (Henry Ford Museum & Greenfield Village, 1989): essays that examine Edison's work at the Menlo Park laboratory. Andre Millard's essay, "Machine Shop Culture and Menlo Park," is particularly important for understanding Edison's approach to innovation in the 1870s.

Michael B. Schiffer, *Taking Charge: The Electric Automobile in America* (Smithsonian Institution Press, 1994): an account of Edison's efforts to develop the storage battery for electric vehicles.

Tom Standage, *The Victorian Internet* (Walker and Company, 1998): an accessible introduction to the development of the telegraph industry in the nineteenth century.

Randall Stross, *The Wizard of Menlo Park: How Thomas Alva Edison Invented the Modern World* (Crown Publishers, 2007): a readable account of Edison as a cultural figure.

Byron M. Vanderbilt, *Thomas Edison, Chemist* (American Chemical Society, 1971): the chemical aspects of Edison inventions.

Michele Wehrwein Albion, *The Quotable Edison* (University Press of Florida, 2011): quotes that trace Edison's relationship with the print media.

Michele Wehrwein Albion, *The Florida Life of Thomas Edison* (University Press of Florida, 2008); and Edward Wirth, *Thomas Edison in West Orange* (Arcadia Publishing, 2008): Edison's relationships with West Orange and Fort Myers.

EDISON LABORATORY
Thomas A. Edison
WEST ORANGE, N.J.

ACKNOWLEDGMENTS

LIKE MOST OF EDISON'S INVENTIONS, this book was a collaborative effort. Many talented people helped edit and proofread text, photograph artifacts, research historic photographs, check facts, and answer questions—among other things—and it's my pleasure to thank them for their assistance.

Michael Fragnito and his team at Sterling Publishing—including editor Melanie Madden, photographer Chris Bain, photography editor Melissa McKoy, copyeditor Joelle Herr, designer Yeon Kim, and creative director Jeff Batzli—turned a rough draft and stacks of historic photos, drawings, and advertisements into this beautiful book.

This book began under former Thomas Edison National Historical Park superintendent Greg Marshall and was completed under his successor, Jill Hawk. I thank them both for their unwavering support.

Assistant Superintendent Teresa Jung, Supervisory Museum Curator Michelle Ortwein, and Chief of Interpretation and Visitor Services Karen Sloat-Olsen cheerfully answered many questions and offered valuable advice about the book's themes and goals.

I also thank my colleagues Jerry Fabris, Joan Harris-Rico, and Holly Marino for making it possible to photograph the museum artifacts presented in this book. Conversations with other colleagues at the park, including Brigid Jennings, Charly Magale, Shemaine McKelvin, Tim Pagano, Carmen Pantaleo, Ed Wirth, and Beth Miller sharpened my understanding of Edison.

Thomas A. Edison Papers Project director Paul Israel helped me appreciate several key points about Edison as an innovator and his role in the development of motion pictures. Chris Pendleton and Alison Giesen of the Edison-Ford Winter Estates shared valuable unpublished information about Edison's rubber research in Florida.

The support of Christine D'Amico, Lorena Lalicata, Harry Roman, and Tom Ungerland of the Charles Edison Fund and Edison Innovation Foundation were also instrumental to the success of this project. I am especially grateful to Foundation president John Keegan for inviting me to work on this book.

Finally, I thank my wife, Anne Markham DeGraaf, and my family for their support.

PHOTO CREDITS

Unless otherwise specified, all images and photographs have been provided by the National Park Service, Thomas Edison National Historical Park.

Alamy: © M. Timothy O'Keefe, 229

Courtesy of Ames Historical Society: xxiv (top)

Art Resource, NY: The New York Public Library, 5; Photograph © SSPL/Science Museum, 31

Christopher Bain: i, v (phonograph), vi (camera box, battery, bulb), vii, xi, xiii, xxviii–1, 10, 16 (bottom), 27 (top), 32-33, 35 (right), 46-47, 54, 68-69, 77 (bottom), 81, 82–83, 84, 85 (top, bottom), 88–89, 91, 95, 96–97, 102, 106, 109 (bottom), 110, 116–117, 122, 128, 179, 211 (iron, toaster, coffee maker), 225, 238

Courtesy of the Bostonian Society: 2

Courtesy of Edison Birthplace Museum, Milan, Ohio: xxiii

Courtesy Edison & Ford Winter Estates, Inc., Fort Myers, FL, www.edisonfordwinterestates.org: 213

Getty Images: Achille-Louis Martinet/The Bridgeman Art Library, 62–63; Mondadori Portfolio, 59

The Granger Collection: 48 (left)

From the collections of The Henry Ford: 18–19, 227 (top)

© Gilbert King: 72, 226

Courtesy of the Lehigh University: 164

Library of Congress: viii (poster), xxv, xxvi–xxvii (top), 14, 43, 90, 108, 123, 124 (top, bottom), 131, 135, 140 (poster), 198 (right); Carol M. Highsmith Archive, 204–205

National Library of Medicine: 158 (bottom)

Courtesy of the National Museum of the U.S. Air Force: 198 (left)

Private Collection: 39

Science Source: Sheila Terry/Science Photo Library, 48 (right)

Courtesy Wikimedia Foundation: xxvii (bottom); David Rumsey Map Collection, xxv; Félix Nadar, 37; Stengel & Co., 65; © Nicholas A. Tonelli, 165; Eberhard J. Wormer, 126

INDEX

Note: Abbreviation "TE" stands for Thomas Edison.